Baris Bagci

Programming and use of TMS320F281 converters

Dokument Nr. V24105 aus dem GRIN Verlagsprogramm

Baris Bagci

Programming and use of TMS320F2812 DSP to control and regulate power electronic converters

GRIN Verlag

Bibliografische Information der Deutschen Nationalbibliothek: Die Deutsche Bibliothek verzeichnet diese Publikation in der Deutschen Nationalbibliografie; detaillierte bibliografische Daten sind im Internet über http://dnb.d-nb.de/ abrufbar.

1. Auflage 2003

http://www.grin.com/
Druck und Bindung: Books on Demand GmbH, Norderstedt Germany
ISBN 978-3-638-70184-6

Programming and Use of TMS320F2812 DSP to Control and Regulate Power Electronic Converters

by

Baris Bagci

Thesis submitted to the Faculty of
Information, Media and Electrical Engineering
in partial fulfillment of the requirements for the degree of

Master of Science
in
Electrical Engineering

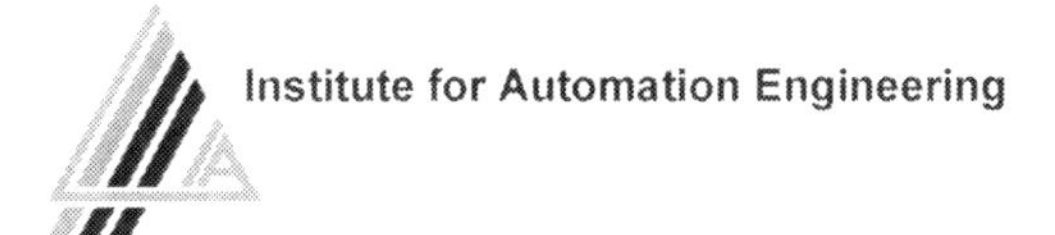

Cologne, October 2003

Abstract

The purpose of this master thesis project has been to study, operate and program the 32-bit 150MIPS TMS320F2812 DSP developed by Texas Instruments Inc. In addition, it has also been a goal to implement fast estimation techniques for control of resonant converters. For this purpose, PWM signals that are generated using this DSP are used. The demands on the system and the hardware to solve the problem were already decided when I started the work.

The algorithms were programmed in C/C++ language, compiled, debugged and transferred to the DSP development board in a compiling and simulation tool (downloader), called CCS (Code Composer Studio v2), also provided by Texas Instruments.

In the first chapters of this thesis I give general information about control systems, digital signal processors, digital signal processing and the DSP used in this work. The following chapters tell about PWM, how to configure the PWM outputs and some examples related with PWM signals are given. After a short review of series resonant converters, I presented the last example implemented in this project. I conclude with a summary and provide some hints of future work.

Acknowledgements

I would like to thank everyone at the Faculty of Information, Media and Electrical Engineering at the University of Applied Sciences Cologne, who have helped me to accomplish this diploma work. Special thanks go to Professor van der Broeck for the valuable assistance, guidance and encouragement he gave during the work. Many thanks to Professor Große for acting as the second referee.

I also wish to thank Mr. Kellersohn and Mr. Küster for the great co-operation.

Last but not least, I would like to thank my parents for the opportunity to do my degree at all.

Declaration

I declare that this thesis is my own work and has not previously been submitted for any other form of assessment. Information derived from the published or unpublished work of others has been acknowledged in the text and a list of references is given.

24.10.2003

Date

Signature

Table of Contents

List of Figures & Tables

Figures

Tables

1. Introduction

1.1 Power Electronic and Electrical Drive Systems

The market for power electronics and power-electronic-controlled electrical drives is rapidly growing. Often, the application of power electronic and electrical drive (PE&ED) systems requires careful engineering because PE&ED systems are utilized as energy converters or as actuators embedded in larger engineering systems. Control systems are required to obtain the desired characteristics of the PE&ED systems and the entire application. The diversity of applications is reflected in a correspondingly wide range of control methods. Hence, a large variety and a large volume of control systems have to be designed, implemented and tested.

1.1.1 Power Electronic Applications

Power electronics constitutes an electrical engineering system, which is always embedded in a system comprising an electric power supply and an electrical load as depicted in Figure 1.1. Note that power electronics itself does not constitute a source of electric power but transforms electrical energy. Electrical energy can be supplied by the electric utility grid comprising remote energy sources and transmission devices. Alternatively, it might be supplied by local energy sources, such as solar cells, wind or hydro generators or energy storage elements like electrochemical batteries [1].

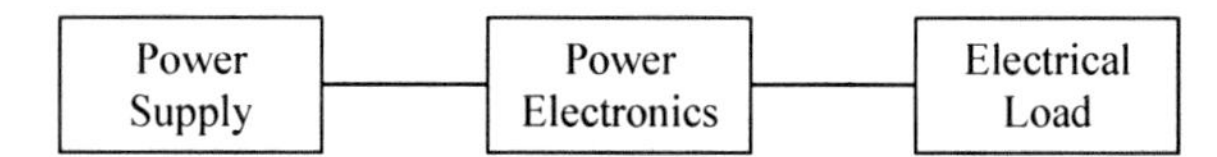

Figure 1.1 Power Electronic System Consisting of Power Electronics and Control Embedded between a Power Supply and an Electrical Load

If the characteristics of the energy source do not meet the requirements of the load then power electronics is used as an interface between the energy source and the load. This might for instance be related to the amplitude of dc systems, the amplitude and frequency of fixed frequency ac systems or disturbances caused by other loads connected to the same power supply. Thus, the task of power electronics can be defined as converting electric energy and controlling the flow of electric power by means of power semiconductor devices in accordance with the requirements of the load. The latter may comprise specifications on amplitude, frequency, and number of phases, allowed harmonic distortion and transient voltages. High efficiency is most often a major requirement for power electronic circuits due to the cost of energy and cooling systems.

Power electronics covers a wide range of applications including the following [1]:

- Energy conversion as part of electrical drives
- DC/DC converters of virtual any power range
- Unity-power-factor rectifiers for applications including railway traction drives which operate of a single-phase ac catenary supply
- Generation of (multi-phase) voltage systems of virtually any frequency or amplitude
- Power engineering applications such as 'flexible ac transmission systems' optimizing the utilization of electrical power transmission equipment

Most current power electronics (PE) systems are implemented using an intermediate dc link between two stages of power conversion. Depending on the energy storage element used, these systems impress voltages or currents on their loads and are thus called voltage source inverters (VSIs) or current source inverters (CSIs), respectively.

1.1.2 Switched Mode Operation

'Switched mode operation' is applied by power electronic circuits due to efficiency requirements and feasibility. In contrast, linear electronic systems use semiconductor devices as adjustable resistors by operating them in their linear region. This results in a low energy efficiency, which is not tolerable or even feasible at high energy levels or the energy density prevailing in power electronics [2]. For this reason, power electronics embodies power semiconductors used as switches that are either (ideally) fully on or fully off. The power semiconductors are operated that way that they switch between conducting and blocking states in a cyclic manner. The high efficiency achieved by this 'switched mode' operation is quite important due to the cost of wasted energy and the difficulty of removing the generated heat.

Most applications of switched mode operation involve the control of mean values to achieve the desired characteristics. Hence, the ratio of the on-state time of a semiconductor to the total cycle time is controlled. It is an important issue in control of power electronics to maximize the control bandwidth despite a limited switching frequency. The switching frequency usually has to be limited because the allowable total losses of the semiconductor devices have to be restricted. Besides off-state leakage and on-state conduction losses, the losses are related to switching losses caused by non-zero voltage and current during the finite duration of the switching operation. The total losses are most often limited by the heat dissipation capability of the package and the permissible power semiconductor junction temperature. A finite switching frequency causes VSI load currents or CSI load voltages to oscillate around the desired mean values.

1.1.3 Electrical Drive Applications

An electrical drive (ED) can be defined as a controlled electromechanical actuator for the purpose of 'motion control' of a larger mechanical engineering plant as shown in Figure 1.2. It consists of the electrical machine, the power converter, the control equipment and the mechanical load [3]. In all drives where the speed and position are controlled, a power electronic converter is needed as an interface between the input power and the motor [4].

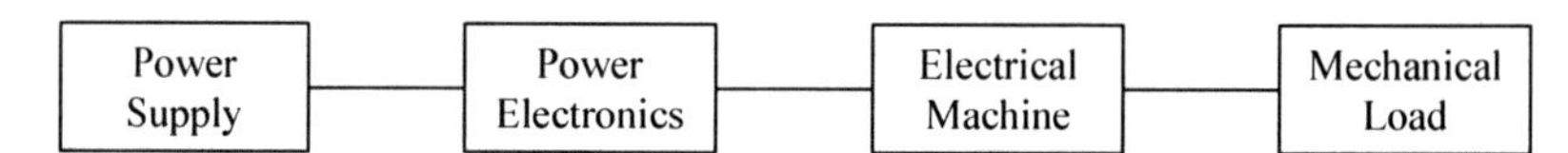

Figure 1.2 Electrical Drive Consisting of Power Electronics, Electrical Machine and Control Embedded between a Power Supply and a Mechanical Load

It has to be emphasized that power electronic is used as tool to provide electric energy to electromechanical actuators for their movement and to achieve control of behavior of the actuators, thus achieving control of the behavior of the mechanical load. Thereby, the characteristics of the power supply to the electrical machine are used to control motion.

Examples of variable speed drive or motion control applications are:

- variable-speed drives for household appliances or industrial purposes
- high-performance servo drives for factory automation
- railroad traction drives
- maritime electrical propulsion systems

1.1.3.1 Motion Control

Motion control deals with the use of high performance electric motors and is a very important part of industrial control systems. Motion control includes applications for speed and torque or position control in practically all branches of industry. An important advance in this field has been made during the last years by the introduction of microprocessor control systems. These systems are becoming a standard in motion control because of fast advances in microelectronics technology and well-known benefits, such as: [5,6,7]

- Greater accuracy
- Flexibility
- Repeatability
- Less noise
- Parameter sensitivity
- Higher interconnection capacity

As a consequence of this progress, more and more applications that have used simple electric drives for economic reasons are being replaced by motion control systems. This is a natural evolution considering the benefits offered by motion control systems in meeting the ever-increasing needs for improved quality and greater productivity in all industries.

To facilitate this move towards motion control solutions, efforts are being made to continuously decrease the cost of these systems, especially considering their power electronics and control parts.

Motion Control System Requirements

Figure 1.3 presents the basic structure of a typical motion control system, which consists of:

- Electric motor (M)
- Power electronic block (PE)
- Digital control system (DCS)

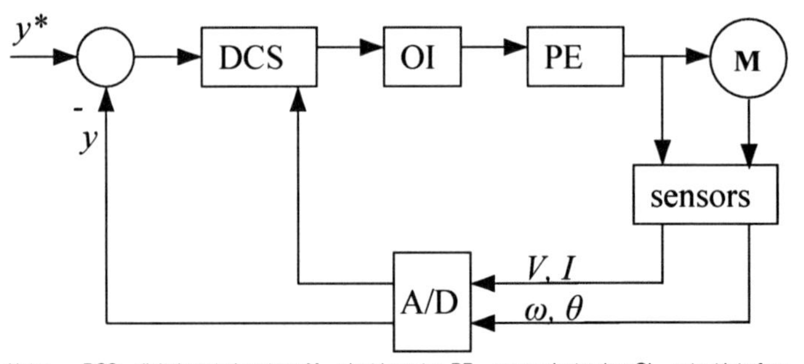

Note: DCS – digital control system; M – electric motor; PE – power electronics; OI – output interface

Figure 1.3 Basic Structure of a Typical Motion Control System

The inputs to the digital controller are the reference quantity y and the process values, which are measured with a set of sensors and generally converted using analogue-to-digital (A/D) converters. These values are electrical (voltage V, current I) and mechanical (velocity w, position q) signals.

The DCS controls the PE through an output interface (OI). This interface consists of digital to analogue (D/A) converters, if the PE requires an analogue input, or of a simple isolation module, if the DCS directly controls the PE switching elements (transistors, IGBTs, etc.).

The requirements for such a system are demanding regarding steady state accuracy and dynamic behavior, sensitivity to external disturbances, measurement noise and parameter variation, and especially reliability and cost.

Inevitably, more and more quantities must be controlled simultaneously (position and/or speed, acceleration, torque, and current). All of these quantities are changing rapidly enough that the sampling time must be very small (typically in the range of 0.1 to 1 ms). Some of these quantities are measured; others can be estimated.

Invariably, there are analog inputs. The number of inputs can range from 5 to 8 for complex systems. This means that the microprocessor system must read the information from the corresponding sensors. Precision depends on the accuracy of the sensors and the converters. Typically these converters must have 10 or more bits of accuracy and a conversion time maximum of 10 ms.

The extreme demands imply that sophisticated control algorithms must be executed in each sampling step. The control scheme of motion control systems is typically complex, even if the controller block is a simple PID (as it is in numerous applications).

For example, in the popular induction motor control principle the complexity derives from the induction motor, which is a nonlinear element. To linearize the induction motor, the equations must be reported to the rotor flux frame. The reference currents obtained in this way must be expressed in the stator frame so that the PE can be controlled. Control diagram complexity differs with the motor type and the system performance.

In many cases the simple PID controller is replaced by more sophisticated schemes, such as adaptive or optimal controllers, to obtain an improved behavior even when parameters change or disturbances occur.

Furthermore, the control scheme can become more complex if quantities necessary for control are estimated rather than measured. In this way the corresponding sensors are eliminated and the system becomes cheaper and more robust.

If analog control is completely eliminated, the digital control system must generate the switching commands for the PE elements. Depending on the motor and PE type, these commands must be computed following a specific algorithm.

For example, the pulse width modulation (PWM) strategy is usually employed for AC motors. The output value can be converted to the corresponding switching sequence either by software or hardware. The hardware solution is more complex but faster.

Aside from the motion control, the microprocessor system must communicate with other systems to receive commands and return process information. This usually requires a special motion control language consisting of a collection of instructions processed by the control system. The microprocessor thus receives the reference not as an array of values but as concise information used to construct an exact profile.

A DSP is frequently used to accomplish such a large number of computations in the small sampling time (0.1 to 1 ms) provided. The DSP solution offers sophisticated control schemes with very good performance.

Because cost is an important restriction for most industrial applications, the scheme must be carefully chosen. An optimum solution must be balanced between the requirement for high performance and the cost of the microprocessor control system.

The cost of a DSP control system depends not only on the cost of the DSP but also the cost of external components (converters, filters, memory, specialized inputs and outputs such as an encoder and a PWM) [8].

A strong tendency in the field of motion control is to integrate the motor, the power converter, and the control electronics in a single, compact unit. This is possible because the size of power electronics is constantly being reduced. An intelligent motor is thus obtained with an extremely simple implementation, increased reliability, and reduced cost [9].

1.2 Control Systems

Control intends to impose certain behavior on an engineering system (plant) in order to satisfy control criteria such as disturbance rejection, low steady-state error, fast transient response and robustness to parameter changes in the plant etc [10]. Control is used to ensure and increase the quality of the performance of the plant. The implementation of a control system or the emulation of a control design by rapid prototyping methods requires appropriate hardware structures.

1.2.1 Digital versus Analog Implementation

In power electronics systems, output voltage regulation has traditionally been accomplished using analog control concepts. In the analog controller, the analog signals of the output voltage and/or current are first processed by an analog transfer function using an operational amplifier with its appropriate compensation network. In this way, the transfer function applies the control laws to shape the output response of the switching converter. The analog controller then adjusts the duty cycle or switching frequency to drive the power switching devices. The main advantage of the analog control system is that the system operates in real time and can have very high bandwidth. Also, the voltage resolution of an analog system is theoretically infinite. However, an analog system is usually composed of hardware that does not lend itself to design changes, and advanced control techniques used to improve performance require an excessive number of analog components.

Even though dedicated analog integrated circuits remain the workhorse of controllers for switching converters, digital controllers are finding more and more applications. It is not only the steady price reduction that has made them attractive in various applications, but also the great functional developments of microcontrollers and digital signal processors (DSPs). As shown in Figures 1.4 and 1.5, the digital controller accepts the digitized sampled output voltage and current through analog-to-digital converters (ADCs). This information is then processed using a digital control algorithm, which is similar to the analog compensation network in terms of function. The output of the digital controller is then used to drive the power switching devices. In general, digital control offers some advantages over the analog counterpart. Some of the advantages are as follows:

- Digital components are less susceptible to aging and environmental variations.
- They are less sensitive to noise.
- Changing a controller does not require an alteration in the hardware.
- They can provide monitoring, self-diagnostics, and communication with a host computer or among several digital controllers.
- The most important is they can facilitate some advanced control techniques, such as space vector modulation, adaptive control, fuzzy control, etc.

However, digital control systems are not without disadvantages when compared with analog control systems. Some of the disadvantages are as follows:

- Finite signal resolution due to the finite wordlength of the ADCs, and DACs or PWM outputs, which cause the output less accurate. Analog control gives infinite resolution of the measured signal.
- Time delays in the control loop due to both the sampling of ADCs and computation of the control algorithm by the processors.
- Analog control can provide continuous processing of signal, thus allowing very high bandwidth.
- It may still need some analog interface circuits, increasing the overall system complexity and cost. However, this problem can be alleviated with more functions integrated into the controller or even application specific integrated circuits (ASICs).

Actually, DSPs or Microcontrollers are already widely used in AC motor drives and uninterruptible power supplies (UPSs), which are both inverter-based systems usually working at a few tens of kHz. They have permitted the application of advanced control techniques in these systems, which is hard to implement with analog control otherwise.

A digital AC drive consists of implementation of torque control by means of regulating the motor current which requires very high speed computation in the range of tens of microseconds, and speed control which requires relatively moderate computation intensity in the range of hundreds microseconds. All these functions have been implemented in one DSP or one microcontroller with/without a separate motion ASIC. This popularity has been due partly to availability and flexibility of desired algorithm implementation.

1.2.1.1 Review of Today's Servo Drive Systems

Today's most servo motor drive systems are implemented by digital closed loop control instead of analog control. This has been primarily due to rapid advancement of Digital Signal Processor (DSP) and microcontrollers applied to motor control applications. In a typical servo control system, several functions are divided into tasks which run at different update rates depending on the required bandwidth and nature of the processing priority need – real-time operation versus delayed batch processes, scanned tasks versus one-time event driven tasks. Each task is controlled by a multitask operating system closely coupled with DSP or microcontroller interrupt structure.

A servo drive system in terms of a functional element, which deals with much closer machine control, requires fast processing, fast update rate and real-time process. They are closely tied with a specific motion peripheral hardware and it sometimes requires specific coding unique to peripheral hardware and interrupt structure inside of DSP or microcontroller.

On the contrary, tasks that are far apart from the machine side and are close to the host communication or man-machine interface side, require less frequent update and slow processing. However, it requires more memory intensive calculation since reference command generation over controlling parameter is more complicated than those, which are close to the machine side. For example, position reference command is much more complex as sophisticated motion profile generation advances. However, torque command is produced in a simple step function.

The fact is that torque is a fastest machine parameter, and needs to be controlled much more quickly than speed of motor shaft. Integral of torque is speed. Integral of speed is position. Integral of power results in motor temperature rise. Because of this chain of physical motor parameters, each parameter requires different speed of processing. It is typical that a real-time multi-tasking operating system is used to satisfy each required processing power [11].

Although digital control has been widely applied to motor drives and UPS applications, the digital control of power supplies faces slightly different technical challenges. In the case of motor drives, the controlled variable of interest is a mechanical quantity, such as position or velocity. Electrical dynamics are often included in the overall design model to achieve high performance goals, but the dominant time constants are associated with mechanical dynamics and are relatively large. Sampling periods are often on the order of several milliseconds to tens of milliseconds, so it is relatively easy to complete control calculations within a sampling period. In addition, the switching frequency is also about one order of magnitude lower than that of most power supplies. Therefore, it is also relatively easy to implement real time PWM control with the-state-of-art digital processors.

In the contrast, power supply applications focus on control of an electrical quantity, such as output voltage. Objectives often include excellent rejection of input and load of variations. The time constants of interest are often several orders of magnitude smaller than for motor drives. Hence, the higher sampling frequency is of greater concern.

Therefore, analog control concept is still the workhorse of most DC/DC converters. For most applications, especially low power DC/DC power supplies, analog control, which is usually realized by a single PWM control chip, still have advantages in terms of cost and simplicity. However, as the applications of power electronics are getting broader and the power electronics systems themselves more complex, the complexity of the system is beyond the capacity of analog control or the performance hardly satisfactory in some applications, e.g., some battery chargers, automotive HID ballast, voltage regulator modules (VRMs).

A microcontroller can be very inexpensive and possess many I/O functions (such as ADCs, timers, etc.), but its computing performance is often significantly lower than a typical DSP. For example, microcontroller may not support a fast multiplication or division instruction not to mention the shorter wordlength of the processor and longer machine cycle even for the same clock rate. Though DSP usually needs more supporting I/O chips, it has the potential of realizing high performance system even compared with analog control. There is also a tendency of integrating more I/O functions into DSPs.

As mentioned earlier, there are some major limitations for digital controller implementation of power converters. The first is output inaccuracy due to the finite wordlength of the ADCs, and DACs or PWM outputs. The second is achievable system bandwidth associated with the sampling and computing delays. The third one is the overall functional integration, which is closely related to the overall system structure and cost.

So for practical purposes, DSP-based system can achieve real time PWM control, even cycle-by-cycle control with satisfactory performance (as shown in Figure 1.5). While microcontroller-based systems have to rely on additional PWM control chips to accomplish the PWM switching function (for a few hundreds of kilohertz) and as an interface a digital- to-analog converter (DAC) is needed here (as shown in Figure 1.4) [12]. The PWM control is far from cycle-by-cycle. The control output from the controller will be updated at a relatively low speed compared to switching frequency and it is a sample and hold process in principle.

So from speed and accuracy point of view, DSP can achieve real-time PWM control up to a few hundreds of kilohertz without additional DAC and PWM control chip and microcontroller cannot and have to rely on additional DAC and PWM control chip for present-day power supplies. It should be noted that achievable bandwidth really depends on the complexity of the control algorithm and the capacity of the instruction set. We only consider the case of single output voltage regulation. It is also possible to realize parallel operation of multiple converters, e.g. VRMs, with more advanced control strategy. Finally, we have to point out that even DSP-based system cannot realize peak current mode control due to the excessive speed requirement of the current sensing though average current mode control is possible.

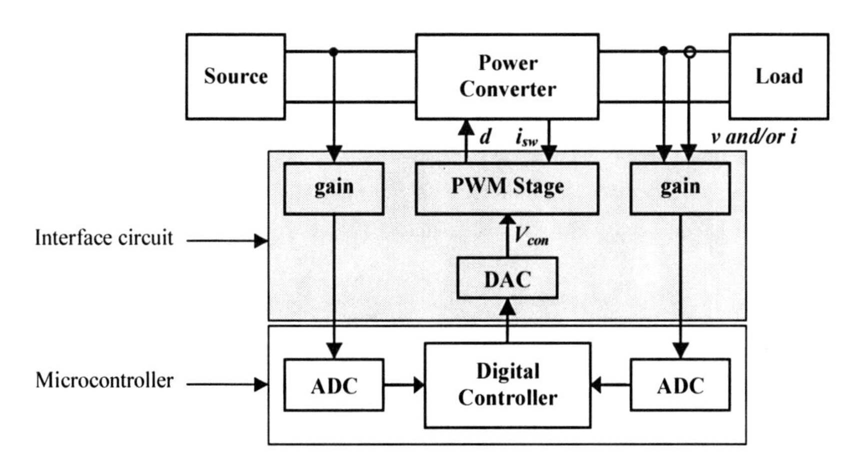

Figure 1.4 Typical Microcontroller-Based Digital Control System Diagram for PWM DC/DC Converter

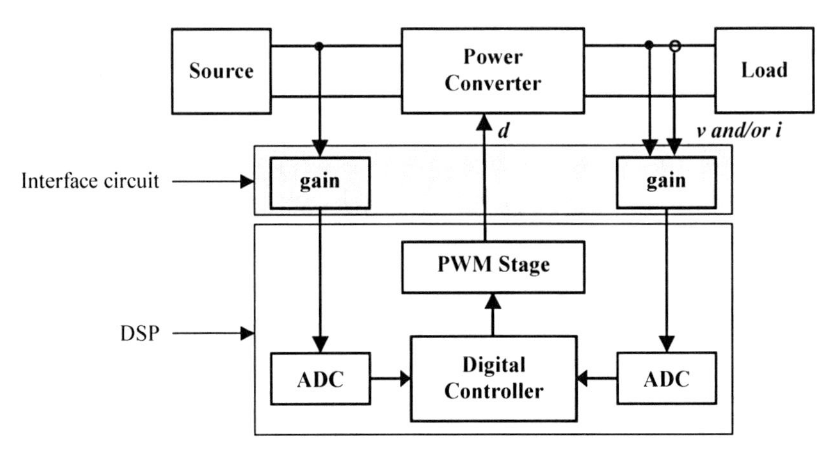

Figure 1.5 Typical DSP-Based Digital Control System Diagram for PWM DC/DC Converter

1.2.2 Digital PWM Control Using DSP

PWM Controllers

Pulse Width Modulation controllers are implemented using both analog and digital control schemes. Pulse width modulator produces a logic signal, which is periodic with frequency *f* and has duty cycle *d*. The signal is used to control the duration over which power transistor in the converter are switched on. The input to the pulse width modulator is an analog control signal. The modulator manipulates the analog control voltage to produce the duty cycle in proportion to it.

With the recent advances in the semi-conductor industry, microprocessors and digital signal processors are used in producing digital PWM. A PWM controller using DSP is shown in Figure 1.6 [13]. In this control scheme the parameters of interest like the output voltage, input voltage and current are fed into a set of preamplifiers and digitized using an A/D converter. These parameters are manipulated suitably in the digital signal processor where the required numerical values namely *Ts*, *Ton*, *Ts*, corresponding to switching period, on-time for the converter switches and start time for the A/D signal conversion are generated. The calculations of these values are based on a set of gain equations. The architecture of this approach is shown in Figure 1.6.

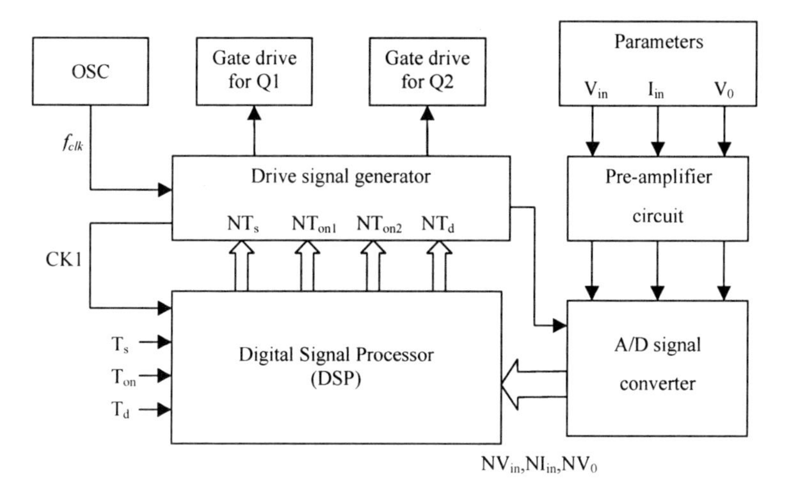

Figure 1.6 Architecture of Digital PWM Using Digital Signal Processor

1.3 Digital Signal Processors

A Digital Signal Processor is a super-fast chip computer, which has been optimized for the detection, processing and generation of real world signals such as voice, video, music, etc, in real time. It is usually implemented in a single chip or nowadays just part of an IC, about 0.5 cm^2 to 4 cm^2 [14].

In contrast, a microprocessor is traditionally a much less powerful computer that performs the mundane tasks, often controlling other devices – e.g. keyboard entry, central heating, washing machine cycles, etc.

Digital signal processors were created many years after general-purpose processors. The need for digital signal processors was due to specific features of digital signal processing algorithms. These algorithms usually include MAC (multiplication and accumulation), where each item of a MAC is in turn a multiplication of other two items. Hence, the main operation in algorithms of digital signal processing is the combination of multiplication and accumulation.

Nearly every DSP is capable of performing a so-called MAC operation in one instruction cycle. The operation consists of a multiply and a subsequent addition (accumulation), as depicted in Figure 1.7.

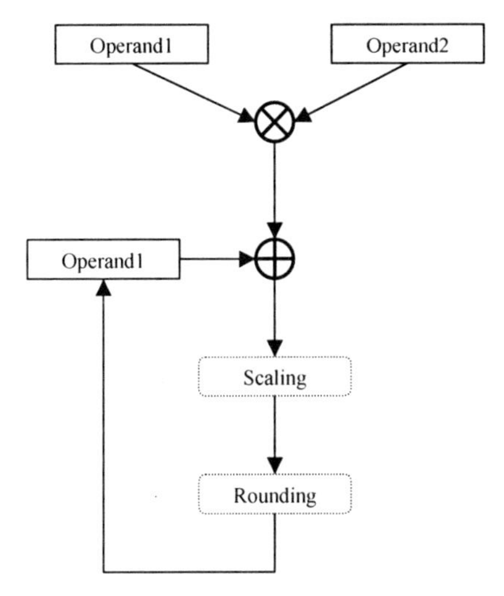

Figure 1.7 MAC Operation

The rapid development of microelectronics makes the manufacturing of very fast digital signal processors easier all the time and also makes it possible to perform more demanding signal processing operations digitally. When compared to general purpose processors, digital signal processors are characterized by the capability of performing the mathematical operations of digital signal processing (DSP) very fast.

The central problem of digital signal processing is to enable systems to work in real time. This means that it is necessary to complete all operations of a signal processing algorithm within the discretization period of this signal. To reach high efficiency in computations mainly based on multiply and accumulate operations in general purpose processors is a difficult task. Therefore, the new type of processors – digital signal processors, which provide an easy and efficient way of realization of DSP algorithms in real time, was offered on market in the early 80s.

DSP is used in all kinds of applications that require digital manipulation of information or digital system control. DSP is especially important to digital cellular telephony and teletechnology in general.

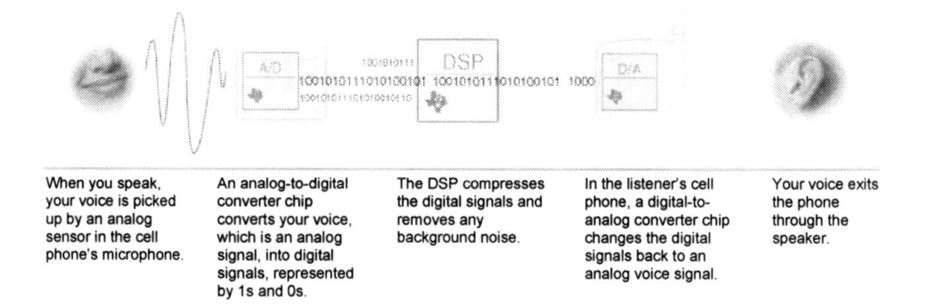

Figure 1.8 In DSPs, an analog signal such as voice is digitized by an analog-to-digital converter (A/D). The DSP processes the digital signal, then a digital-to-analog converter (D/A) changes the signal back to analog [15].

What is actually the difference between digital signal processors and general-purpose processors? To reach the necessary speed in digital signal processors, the main hardwired operation of DSP is the multiply-accumulate operation, which can be performed within one clock cycle. From the architectural point of view digital signal processors use different, so-called modified Harvard processor architecture that enables command pipelining (processing parts of commands in parallel). Memory is divided into two parts: data memory and instruction memory. It is possible to move data directly between these two types of memory. This approach made possible to organize efficient parallelization of calculations, i.e. "fetch, decode, execute" phases of adjacent instructions can be performed in parallel.

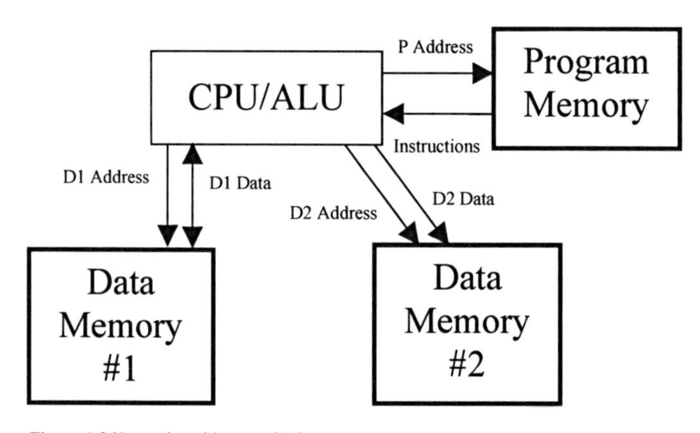

Figure 1.9 Harvard Architecture [16]

Conventional microprocessors use the Von Neumann architecture: program and data all in a single memory. Address and data buses are shared between instruction and data fetches [16].

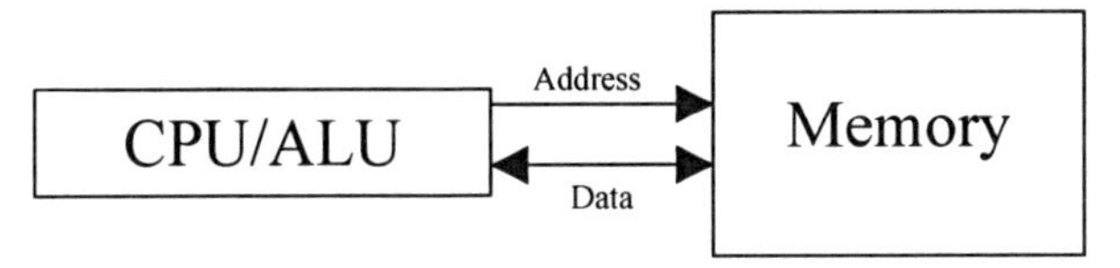

Figure 1.10 Von Neumann Architecture

While digital signal processors are well suitable for signal processing, general-purpose processors are better on general computing applications. Desktop general-purpose processors have hardware support for floating point and integer arithmetic.

Often digital signal processors are divided into two types, floating-point processors and fixed-point processors, depending on the type of arithmetic used in the processor for calculations. Digital signal processors can also be divided into algorithm-specific and application-specific processors. Desktop general-purpose processors must be able to support code written for previous designs of these processors. They must support object-code compatibility. In case of digital signal processors it is not so critical. Digital signal processors and general-purpose processors also differ in memory organization principles. For general-purpose processors it is traditional to use on-chip instruction and data caches as their only on-chip RAM. They also support off-chip memory and second-level caches. Digital signal processors usually have on-chip RAM and very seldom use caches.

Usually most digital signal processors share several common features to support high-performance, repetitive, numerically intensive tasks. As already mentioned above, one of the central operations of most digital signal processors is the multiply and accumulate operation (MAC). All digital signal processors have the ability to perform multiply-accumulate operations in a single instruction cycle. To support such a feature, digital signal processors include accumulator and multiplier integrated into a special arithmetic processing unit in the data path of the processor. Some modern digital signal processors provide two or even more multiply-accumulate units. This feature allows multiply-accumulate operations to be performed in parallel. Generally digital signal processors provide extra "guard" bits in the accumulator to allow a series of multiply-accumulate operations to proceed without causing arithmetic overflow (when numbers exceed the maximum value the processor's accumulator can hold).

A beneficial feature shared by digital signal processors is called multiple-access memory architecture. This means that it is common for most digital signal processors to provide the ability to complete several accesses to memory in a single instruction cycle. This feature allows the processor to fetch an instruction while simultaneously fetching operands and storing the result of a previous instruction to memory. For instance, for computing the vector dot product for an FIR filter, most digital signal processors are able to perform a MAC while simultaneously loading the data sample and coefficient for the next MAC. But on the other hand such single-cycle multiple memory accesses are often subject to some limitations. Usually, all but one of the memory locations accessed must reside on-chip, and multiple memory accesses can only take place with certain instructions. Digital signal processors provide multi-ported on-chip memories, multiple on-chip buses, and sometimes multiple independent memory banks to support simultaneous access of multiple memory locations.

Another feature often used to speed up arithmetic processing on digital signal processors is one or more dedicated address-generating units. This feature enables performing address calculation for operand access in parallel with the execution of arithmetic instructions. General-purpose processors usually require additional cycles to generate the addresses needed to load operands. Address generation units of digital signal processors usually support several addressing modes. These modes are register-indirect addressing with post increment/ decrement (for example, this mode is widely used during repetitive computations on data stored sequentially in memory). Another mode, called modulo addressing, is often supported to simplify the use of circular buffers. Some processors also support bit-reversed addressing, which increases the speed of fast Fourier (FFT) algorithms.

Many DSP algorithms involve performing repetitive computations. Because of this fact, most DSP processors provide special support for efficient looping. Special repeat or loop instruction is provided, allowing implementation of a for-next loop without spending extra instruction cycles for updating and testing the loop counter or branching back to the top of the loop. It is called zero-overhead looping.

An additional feature shared by digital signal processors is the ability to allow low-cost, high-performance input and output. Most digital signal processors incorporate one or more serial or parallel I/O interfaces, and specialized I/O handling mechanisms such as low-overhead interrupts and direct memory access (DMA), which allow data transfers to proceed with little or no intervention from the rest of the processor [17].

There are a lot of DSP algorithms and many system applications use it. Digital signal processors find wide usage in speech coding/decoding, modem algorithms, image compression spectral estimation, hi-fi audio and so on. We do not describe the details of the general purpose digital signal processor architecture here, because one of the following chapters contains a description of TMSC320F2812 DSP developed by Texas Instruments, which has been used for the purposes of this work.

1.3.1 Data Path of a DSP

The data path of a digital signal processor is where the vital arithmetic manipulations of signals take place. DSP data paths are highly specialized to achieve high performance on the types of computation most common in DSP applications, such as multiply-accumulate operations. Registers, adders, multipliers, comparators, logic operators, multiplexers, and buffers represent 95% of a typical DSP data path.

Multiplier

A single-cycle multiplier is the essence of a DSP since multiplication is an essential operation in all DSP applications. An important distinction between multipliers in DSPs is the size of the product according to the size of the operands. In general, multiplying two n-bit fixed-point numbers requires a 2xn bits to represent the correct result. For this reason DSPs have in general a multiplier, which is twice the word length of the native operands.

Accumulators & Registers

Accumulators and registers hold intermediate and final results of multiply-accumulate and other arithmetic operations. Most DSP processors have two or more accumulators. In general, the size of the accumulator is larger than the size of the result of a product. These additional bits are called guard bits. These bits allow accumulating values without the risk of overflow and without rescaling. N additional bits allow up to 2^n accumulations to be performed without overflow. Guard bits method is more advantageous than scaling the multiplier product since it allows the maximum precision to be retained in intermediate steps of computations.

ALU

Arithmetic logic units implement basic arithmetic and logical operations. Operations such as addition and subtraction are performed in the ALU.

Shifter

In fixed-point arithmetic, multiplications and accumulations often induce a growth in the bit width of results. Scaling is then necessary to pass results from stage to stage and is performed through the use of shifters.

1.3.2 Peripherals of a DSP

Most digital signal processors provide on-chip peripherals and interfaces to allow the DSP to be used in an embedded system with a minimum amount of external hardware to support its operation and interfacing.

Serial Port

A serial interface transmits and receives data one bit at a time. These ports have a variety of applications like sending and receiving data samples to and from A/D and D/A converters and codecs, sending and receiving data to and from other microprocessors or DSPs, communicating with other hardware. The two main categories are synchronous and asynchronous interface. The synchronous serial ports transmit a bit clock signal in addition to the serial bits. The receiver uses this clock to decide when to sample received data. On the opposite, asynchronous serial interfaces do not transmit a separate clock signal; they rely on the receiver deducing a clock signal from the data itself.

Direct extension of serial interfaces leads to parallel ports where data are transmitted in parallel instead of sequentially. Faster communication is obtained through costly additional pins.

Host Port

Some DSPs provide a host port for connection to a general-purpose processor or another DSP. Host ports are usually specialized 8 or 16 bit bi-directional parallel ports that can be used to transfer data between the DSP and the host processor.

Link Port or Communication Port

This kind of port is dedicated to multiprocessor operations. It is in general a parallel port intended for communication between the same types of DSPs.

Interrupt Controller

An interrupt is an external event that causes the processor to stop executing its current program and branch to a special block of code called an interrupt service routine. Typically this code deals with the origin of the interrupt and then returns from the interrupt. There are different interrupt sources: on-chip peripherals: serial ports, timers, DMA,...
External interrupt lines: dedicated pins on the chip to be asserted by external circuitry

Software interrupts: also called exceptions or traps, these interrupts are generated under software control or occurs for example for floating-point exceptions (division-by-zero, overflow and so on).

DSPs associate interrupts with different memory locations. These locations are called interrupt vectors. These vectors contain the address of the interrupt routines. When an interrupt occurs, the following scenario is often encountered:

- Save program counter in a stack
- Branch to the relevant address given by the interrupt vector table
- Save all registers used in the interrupt routine
- Perform dedicated operations

- Restore all registers
- Restore program counter

Priority levels can be assigned to the different interrupt through the use of dedicated registers. An interrupt is acknowledged when its priority level is strictly higher that current priority level.

Timers

Programmable timers are often used as a source of periodic interrupts. Completely software-controlled to activate specific tasks at chosen times. It is generally a counter that is preloaded with a desired value and decremented on clock cycles. When zero is reached, an interrupt is issued.

DMA

Direct Memory Access is a technique whereby data can be transferred to or from the processor's memory without the involvement of the processor itself. DMA is commonly used to provide improved performance with input/output devices. Rather than have the processor read data from an I/O device and copy the data into memory or vice versa, a separate DMA controller can handle such transfers in parallel.

1.4 Digital Signal Processing

To 'digital signal process' is to manipulate signals that have either originated in, or are to be exported to, the real world, where those signals are represented as digits (numbers) [14]. Application of digital processing devices to transmit, store and process analog signals require their representation in units of discrete points as numerical values. This means that an integral part of DSP applications is the conversion of the real world signals – analog voice, music, video, engine speed, ground vibration – to numerical values for processing by the DSP – a process termed analog to digital conversion (A/D). Discretization can be performed by analog-to-digital converter (ADC) at the input of DSP system. Going in the other direction – the conversion of numerical values generated within the DSP to real world signal – a process termed digital to analog conversion (D/A) is also involved. Restoration of signal back to analog form is made via digital-to-analog converter (DAC) at the output. To digitally represented signals different mathematical operations can be applied and that is the essence of digital signal processing.

Digital signal processing is used in a wide range of applications, in which the nature of processed signals can be very different – radio, sound, video etc. Fundamental to DSP is the fact that any real world signal, e.g. music, can be accurately represented by samples of the signal taken at periodic intervals. These samples can then be converted into numbers (e.g. representing the volume of the music at the sample point), and these numbers are expressed in binary form. The level of processing complexity can vary depending on application specifics and performance requirements.

Why did digital signal processing become so popular? The answer is in fact that digital signal processing has several advantages compared to analog signal processing. DSP systems are able to complete the tasks inexpensively so to say. Especially it refers to control part of applications. It is much more difficult to control analog signals processing. First of all, analog systems are more sensitive to environment conditions. For example, analog circuit performance can depend a lot on its temperature and change behavior according to changes in its temperature. Analog components have certain tolerances in their values. That means that two circuits of the same design can respond differently having little differences in their internal analog components. In contrast, two digital systems with identical functionality will produce the same results on the same input. This makes digital systems more predictable and reliable than analog.

DSP systems are reprogrammable. One DSP system can be programmed to perform different tasks. If you want to change functionality on analog system - other physical components should be installed. Analog components tend to have different size for different values. For example to increase capacitance, bigger capacitor should be used. In DSP processing you would need only to increase the value in the program, but the hardware part stays unchanged.

Depending on class of applications different aspects are important. The largest class of DSP applications is inexpensive high-volume embedded systems, such as modems, cellular telephones, pagers and so on. Here, cost and power consumption are factors of main importance. Another class is processing large amounts of data using complex algorithms, like in sonars, radars and seismic exploration. Performance and multiprocessor support are favorites among aspects for these applications. A separate category consists of Personal Computer (PC) -based multimedia applications. It becomes very popular to integrate DSP systems into PCs to perform multimedia functions, like voice mail or music synthesis Performance is also important for these devices because the DSP part can be asked to execute several algorithms simultaneously. Memory capacity may be an important issue to consider in these applications too [17].

1.4.1 The History of DSP

Since the invention of the transistor and integrated circuit, digital signal processing functions have been implemented on many hardware platforms ranging from special-purpose architectures to general-purpose computers. DSP as a device, a process or a subject at University is relatively new. In the 1960s, DSP hardware used discrete components and consequently, because of the high cost and volume, its application could only be justified for very specialized requirements. In the 1970s, monolithic components for some of the DSP subsystems appeared, primarily dedicated digital multipliers and address generators, and DSP systems could be implemented using bit slice microprocessors. The breakthrough for mass exploitation of DSP techniques came in 1979 when Intel introduced the 2920, a completely self-contained signal processing device in a 40-pin DIP package incorporating on-board program EPROM, data RAM, A/D and D/A converters, and an architecture and instruction set powerful enough to implement a full duplex 1200 bps modem, including transmit and receive filters. This breakthrough was followed up convincingly by Texas Instruments in 1982 with the launch of the TMS32010.

Since the Intel 2920 there have been five further generations of general-purpose signal processing devices, with a sixth generation announced. The latest devices have some 100000 times the processing power of that early 2920 device, all within the space of 20 years.

The following table summarizes the evolution of DSPs:

Date	Features	Example processors
First generation : 1979 - 1985	Harvard architecture, hardwired multiplier	NECµPD7720, Intel 2920, Bell Labs DSP1, TMS320C10
Second generation: 1985 - 1988	Concurrency, multiple busses, on-chip memory	TMS320C25, MC56001, DSP16 (AT&T)
Third generation: 1988 - 1992	On-chip floating point operations	TMS320C30, MC96002, DSP32C (AT&T),
Fourth generation: 1992 - 1997	Multi-processing features Image and video processors Low-power DSPs (AT&T)	TMS320C40&50, TMS320C80
Fifth generation: 1997 –	VLIW	TMS320C6x, Philips TriMedia, Motorola Starcore

Table 1.1 Evolution of DSPs.

Today, all of the major semiconductor device manufacturers have DSP products either already released or under development, reflecting their confidence in an enormous growth of demand in the late 1980s, 1990s and now into the 21st century, paralleling the spectacular success of microprocessors in the 1970s.

Digital signal processing in general and DSP processors in particular, are used in a wide variety of applications from military radar systems to consumer electronics. Naturally, no one processor can meet the needs of all applications. Criteria such as performance, cost, integration, ease of development, power consumptions are key points to examine when designing or selecting a particular DSP for a class of applications.

Speech analysis was the driving force behind initial attempts to process signals digitally. The exacting tolerances demanded of filters in speech processing systems simply could not be maintained over time with analog techniques, subject as they are to temperature drift, component tolerances, and aging. Digital processing operations, in contrast, consisting of nothing more than sequences of binary multiplication, addition, subtraction, etc, make the outcome entirely predictable and thus reproducible. This feature, together with the ability to replicate multiple analog processing tasks within a single low-power device, is the reason why DSP is a thriving multi-billion dollar industry [14].

2. The TMS320F2812 DSP

2.1 Overview

The TMS320F2812 device, a member of the TMS320C28x. DSP generation, is highly integrated, high-performance solution for demanding control applications and is the first 32-bit 150MIPS DSP with on-chip flash memory and on-chip high-precision analog peripherals. In addition, its architecture is specially optimized for C/C++. Further, this device enables users to develop their code in virtual floating point via the IQ math capability. In this chapter, TMS320F2812 is abbreviated as F2812.

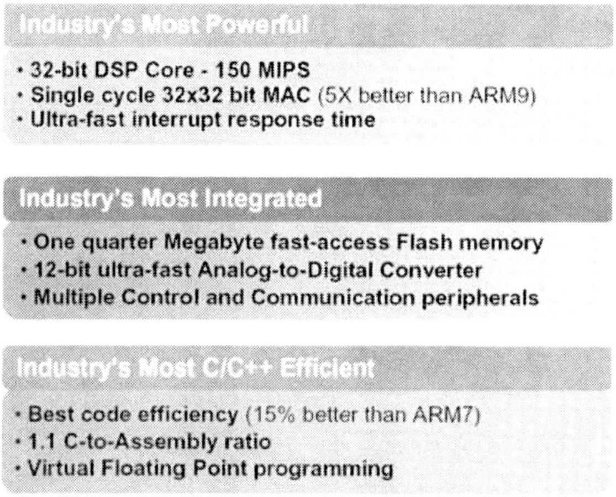

Industry's Most Powerful
• 32-bit DSP Core - 150 MIPS
• Single cycle 32x32 bit MAC (5X better than ARM9)
• Ultra-fast interrupt response time

Industry's Most Integrated
• One quarter Megabyte fast-access Flash memory
• 12-bit ultra-fast Analog-to-Digital Converter
• Multiple Control and Communication peripherals

Industry's Most C/C++ Efficient
• Best code efficiency (15% better than ARM7)
• 1.1 C-to-Assembly ratio
• Virtual Floating Point programming

Figure 2.1 A New TI DSP Product Line: 32-bit Flash Mixed Signal DSP [18]

The embedded control DSP, F2812, is based on a 32 bit DSP core delivering 150 MIPS of performance on a flash process and an impressive 32x32bit MAC in a single 6.67ns cycle. This DSP also feature a large amount of fast-access on-chip flash memory so that code can be executed internally without adding costly external flash memories. In addition, it incorporates a high-precision ultra-fast analog to digital converter (ADC) together with many control and communication peripherals for truly single-chip designs.

The huge majority of embedded applications are not requiring greater than 150 MIPS of performance, greater than 12 bits of analog precision, or an external 16 Mbyte commodity flash memory. As a matter of fact, solutions featuring up to 150 MIPS of DSP performance, a solid 10bit or 12bit Analog to Digital Converter and several kilo-words of flash memory on the same chip, are not only "good enough to do the job" but also provide a much lower cost solution.

The C28x DSP core is designed to have general purpose processor (GPP) features like a unified memory space and addressable registers for faster interrupts and a very efficient C/C++ compiler. Some of the key core features are:

• Architecture designed & optimized for GPP
• Efficient atomic operations those are common in GPP
• Most common instructions that are coded in 16-bits
• C/C++ compiler tuned for GPP

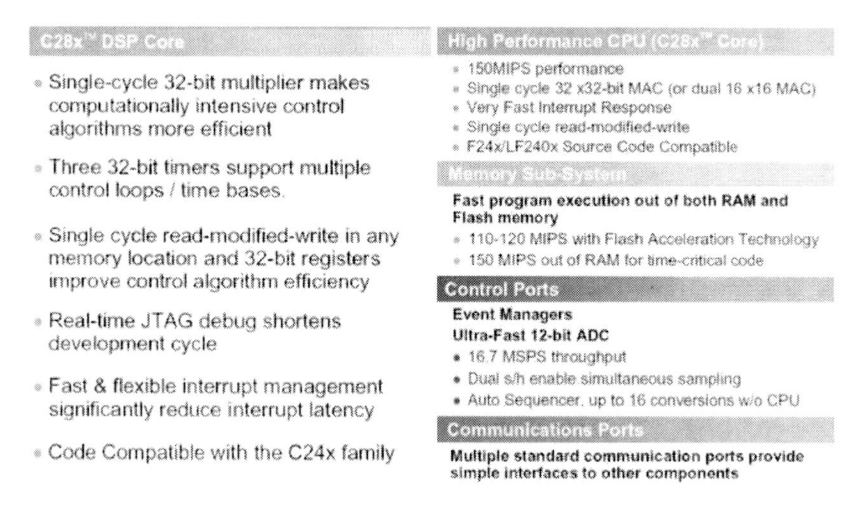

Figure 2.2 C28x DSP Core.

The F2812 feature a very fast integrated 12 bit ADC. This 16 channel high-speed ADC has a single conversion time of 200 ns. However to further increase performance this pipelined ADC is capable of achieving up to 16 conversions at 60 ns each without any CPU intervention.

The ADC also has the ability to auto-sequence a series of conversions. The versatile sequencer can be operated as 2 independent 8-states sequencer or as a large 16-states sequencer [18].

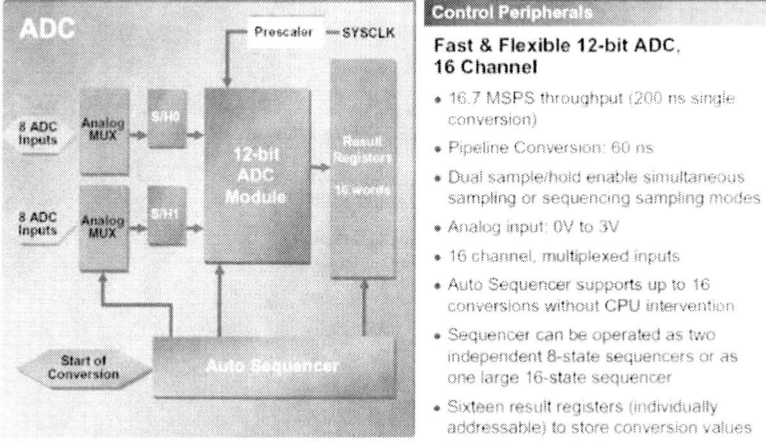

Figure 2.3 On-Chip 12-bit Analog-to-Digital Converter [18].

The F2812 has 2 Event Manager peripherals on-chip to provide a broad range of functions and features that are particularly useful in control applications.

The event manager modules include general-purpose (GP) timers, full-compare/PWM units, capture units, and quadrature-encoder pulse (QEP) circuits. The 2 Event Manager modules are identical peripherals intended for multi-axis/digital control applications.

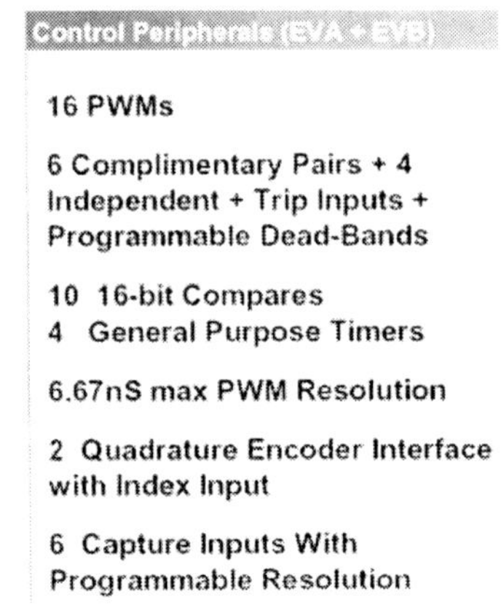

Figure 2.4 2 On-Chip Event Managers.

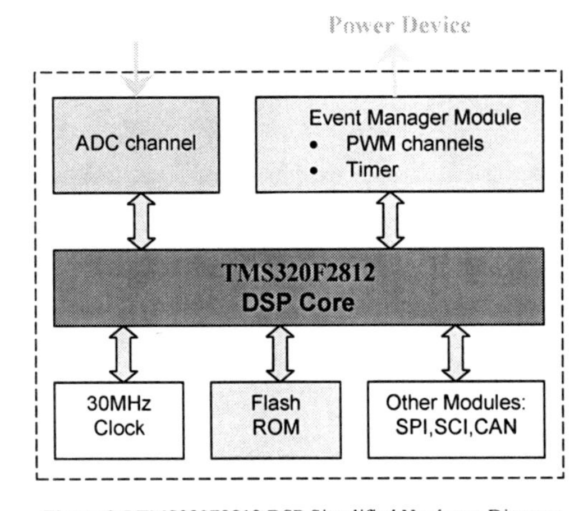

Figure 2.5 TMS320F2812 DSP Simplified Hardware Diagram

The IQ math library was developed to allow the user to start code development in the floating point space. These algorithms are typically written in high level language like C/C++. However, system cost requirements demand that these algorithms are ported to fixed point math [18]. The IQ math library allows the users to code their algorithms in such a way that it seems like they are coding with the ease of floating point, even though this DSP is a fixed point machine.

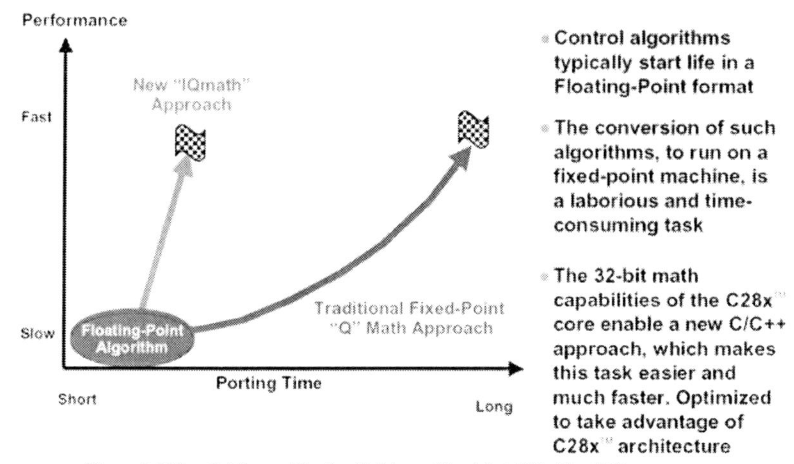

Figure 2.6 IQmath Library: Floating Point on a Fixed Point Machine [18]

The traditional fixed point Q math approach is very time consuming. The combination of TI's C28x 32 bit numerical resolution and IQ math library enable the transition from floating point to fixed point in a matter of minutes.

IQmath is a mathematical approach and a set of supporting math libraries that enable the following:

- Reduced implementation/porting/debugging time of math algorithms in C/C++
- Increased numerical resolution of algorithms from 16-bits to 32/64-bits (near floating-point)

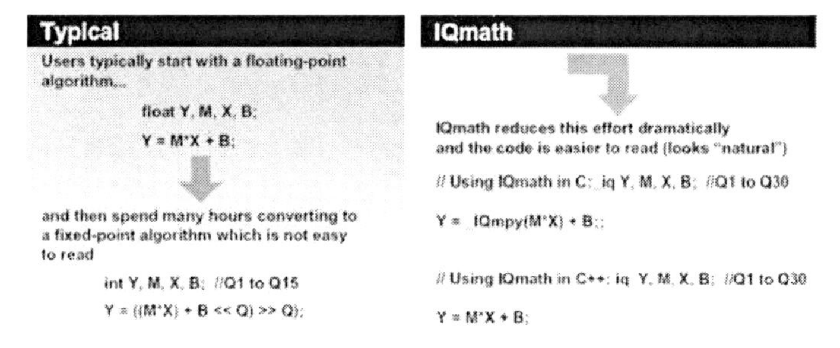

Figure 2.7 IQmath Approach

2.2 The Peripherals of F2812

The peripherals of F2812 are summarized in this chapter. The related ones with this project are described before and continued on the following pages. F2812 has the following peripherals:

- System Control and Interrupts
- External Interface (XINTF)
- Enhanced Controller Area Network (eCAN)
- Multichannel Buffered Serial Port (McBSP)
- Serial Peripheral Interface (SPI)
- Serial Communications Interface (SCI)
- Boot ROM
- Analog-to-Digital Converter (ADC):
- Event Manager (EV)

System Control and Interrupts

System control and Interrupts includes the following modules:

- Memory
- Code security module (CSM)
- Clocking
- General-purpose inputs/outputs (GPIO)
- System configuration
- Peripheral interrupt expansion (PIE)

Code Security Module (CSM)

Security is defined with respect to the access of the on-chip program memory and prevents unauthorized copying of proprietary code. The code security module (CSM) blocks access to on-chip program memory.

Peripheral Clocking

The clocks to each individual peripheral can be enabled/disabled so as to reduce power consumption when a peripheral is not in use. Additionally, the system clock to the serial ports and the event managers, CAP and QEP blocks can be scaled relative to the CPU clock. This enables the timing of peripherals to be decoupled from increasing CPU clock speeds.

Timers

CPU-Timers 0, 1, and 2 are identical 32-bit timers with presettable periods and with 16-bit clock prescaling. The timers have a 32-bit count down register, which generates an interrupt when the counter reaches zero. The counter is decremented at the CPU clock speed divided by the prescale value setting. When the counter reaches zero, it is automatically reloaded with a 32-bit period value.

Watchdog

The F2812 support a watchdog timer. The user software must regularly reset the watchdog counter within a certain time frame; otherwise, the watchdog will generate a reset to the processor. The watchdog can be disabled if necessary.

General-purpose Input/Output (GPIO) Multiplexer

Most of the peripheral signals are multiplexed with general-purpose I/O (GPIO) signals. This enables to use a pin as GPIO if the peripheral signal or function is not used. On reset, all GPIO pins are configured as inputs. One can then individually program each pin for GPIO mode or Peripheral Signal mode.

Peripheral Interrupt Expansion (PIE) Block

The PIE block multiplexes numerous interrupt sources into a smaller set of interrupt inputs. The PIE block can support up to 96 peripheral interrupts. On the F2812, 45 of the possible 96 interrupts are used by peripherals. The interrupts are grouped into blocks of eight and each group is fed into one of 12 CPU interrupt lines (INT1 to INT12). Each of the 96 interrupts is supported by its own vector stored in a dedicated RAM block that can be overwritten by the user. The vector is automatically fetched by the CPU on servicing the interrupt. It takes nine CPU clock cycles to fetch the vector and save critical CPU registers. Therefore, the CPU can respond quickly to interrupt events. Prioritization of interrupts is controlled in hardware and software [19].

Enhanced Controller Area Network (eCAN)

This is the enhanced version of the CAN peripheral. It supports 32 mailboxes, time stamping of messages, and is CAN 2.0B-compliant.

Multichannel Buffered Serial Port (McBSP)

The McBSP is used to connect to E1/T1 lines, phone-quality codecs for modem applications or high-quality stereo-quality Audio DAC devices. The McBSP receive and transmit registers are supported by a 16-level FIFO. This significantly reduces the overhead for servicing this peripheral.

Serial Port Interface (SPI)

The SPI is a high-speed, synchronous serial I/O port that allows a serial bit stream of programmed length (one to sixteen bits) to be shifted into and out of the device at a programmable bit-transfer rate. Normally, the SPI is used for communications between the DSP controller and external peripherals or another processor. Typical applications include external I/O or peripheral expansion through devices such as shift registers, display drivers, and ADCs. Multi-device communications are supported by the master/slave operation of the SPI. On the F2812, the port supports a 16-level, receive and transmit FIFO for reducing servicing overhead [19].

Serial Communications Interface (SCI)

The SCI is a two-wire asynchronous serial port, commonly known as UART. On the F2812, the port supports a 16-level, receive and transmit FIFO for reducing servicing overhead.

Analog-to-Digital Converter

The TMS320F28x. ADC module is a 12-bit pipelined analog-to-digital converter (ADC). It contains two sample-and-hold units for simultaneous sampling. It has 16 channels, configurable as two independent 8-channel modules to service event managers A and B. The two independent 8-channel modules can be cascaded to form a 16-channel module. Although there are multiple input channels and two sequencers, there is only one converter in the ADC module. Figure 2.8 shows the block diagram of the F2812 ADC module [20].

Functions of the ADC module include:

- 12-bit ADC core with built-in dual sample-and-hold (S/H)
- Dual simultaneous sampling or sequential sampling modes
- Analog input: 0 V to 3 V
- Fast conversion time:
- Single conversion time: 200 ns
- Pipelined conversion time: 60 ns
- 16-channel, multiplexed inputs
- Sixteen result registers (individually addressable) to store conversion values
- External pin

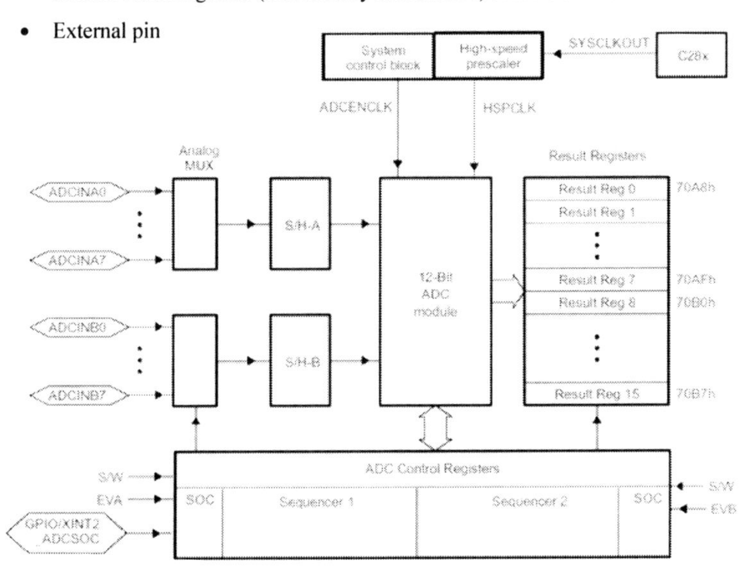

Figure 2.8 Block Diagram of the F2812 ADC Module [20]

To obtain the specified accuracy of the ADC, proper board layout is very critical. To the best extent possible, traces leading to the ADCINxx pins should not run in close proximity to the digital signal paths. This is to minimize switching noise on the digital lines from getting coupled to the ADC inputs. Furthermore, proper isolation techniques must be used to isolate the ADC module power pins from the digital supply.

Event Manager

The event manager (EV) modules provide a broad range of functions and features that are particularly useful in motion control and motor control applications. The event-manager modules include general-purpose (GP) timers, full-compare/PWM units, capture units, and quadrature-encoder pulse (QEP) circuits. Two such event managers are provided, which enable two three-phase motors to be driven or four two-phase motors [19].The two EV modules, EVA and EVB, are identical peripherals, intended for multi-axis/motion-control applications.

General-Purpose (GP) Timers

There are two GP timers. The GP timer x (x = 1 or 2 for EVA; x = 3 or 4 for EVB) includes:

- A 16-bit timer, up-/down-counter, TxCNT, for reads or writes
- A 16-bit timer-compare register, TxCMPR, for reads or writes
- A 16-bit timer-period register, TxPR, for reads or writes
- A 16-bit timer-control register, TxCON, for reads or writes
- A selectable direction input pin (TDIRx) (to count up or down when directional up-/down-count mode is selected)

The GP timers can be operated independently or synchronized with each other. The compare register associated with each GP timer can be used for compare function and PWM-waveform generation. There are three continuous modes of operations for each GP timer in up- or up/down-counting operations. Internal or external input clocks with programmable prescaler are used for each GP timer. GP timers also provide the time base for the other event manager sub modules: GP timer 1 for all the compares and PWM circuits, GP timer 2/1 for the capture units and the quadrature-pulse counting operations. Double-buffering of the period and compare registers allows programmable change of the timer (PWM) period and the compare/PWM pulse width as needed [21].

Full-Compare Units

There are three full-compare units on each event manager. These compare units use GP timer1 as the time base and generate six outputs for compare and PWM-waveform generation using programmable deadband circuit. The state of each of the six outputs is configured independently. The compare registers of the compare units are double-buffered, allowing programmable change of the compare/PWM pulse widths as needed.

Programmable Deadband Generator

The deadband generator circuit includes three 8-bit counters and an 8-bit compare register. Desired deadband values can be programmed into the compare register for the outputs of the three compare units. The deadband generation can be enabled/disabled for each compare unit output individually.

The deadband-generator circuit produces two outputs (with or without deadband zone) for each compare unit output signal. The output states of the deadband generator are configurable and changeable as needed by way of the double-buffered ACTRx register [21].

3. The eZdsp F2812 Board

3.1 Overview

The eZdsp F2812 is a stand-alone card to develop and run software for the TMS320F2812 processor. It is a 5.25 x 3.0 inch, multi-layered printed circuit board, powered by an external 5-Volt only power supply.

The eZdsp F2812 is shipped with a TMS320F2812 DSP. It allows full speed verification of F2812 code. Two expansion connectors are provided for any necessary evaluation circuitry not provided on the as shipped configuration. To simplify code development and shorten debugging time, a C2000 Tools Code Composer driver is provided. In addition, an onboard JTAG connector provides interface to emulators, operating with other debuggers to provide assembly language and 'C' high level language debug [22].

Key Features of the eZdsp F2812

The eZdsp F2812 has the following features:

- TMS320F2812 Digital Signal Processor
- 150 MIPS operating speed
- 18K words on-chip RAM
- 128K words on-chip Flash memory
- 64K words off-chip SRAM memory
- 30 MHz clock
- 2 Expansion Connectors (analog, I/O)
- Onboard IEEE 1149.1 JTAG Controller
- 5-volt only operation with supplied AC adapter
- TI F28xx Code Composer Studio tools driver
- On board IEEE 1149.1 JTAG emulation connector

Functional Overview of the eZdsp F2812

Figure 3.1 shows a block diagram of the basic configuration for the eZdsp F2812. The major interfaces of the eZdsp are the JTAG interface, expansion interface, analog interface and I/O interface.

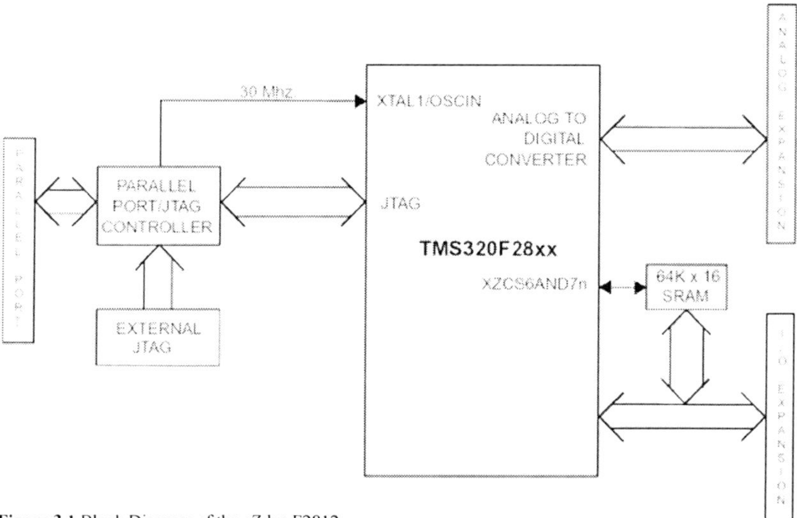

Figure 3.1 Block Diagram of the eZdsp F2812

eZdsp F2812 Memory

The eZdsp includes the following on-chip memory:

• 128K x 16 Flash

• 2 blocks of 4K x 16 single access RAM (SARAM)

• 1 block of 8K x 16 SARAM

• 2 blocks of 1K x 16 SARAM

In addition 64K x 16 off-chip SRAM is provided. The processor on the eZdsp can be configured for boot-loader mode or non-boot-loader mode [22].

3.2 eZdsp F2812 Connectors

The eZdsp F2812 has five connectors. Pin 1 of each connector is identified by a square solder pad. The diagram below shows the position of each connector.

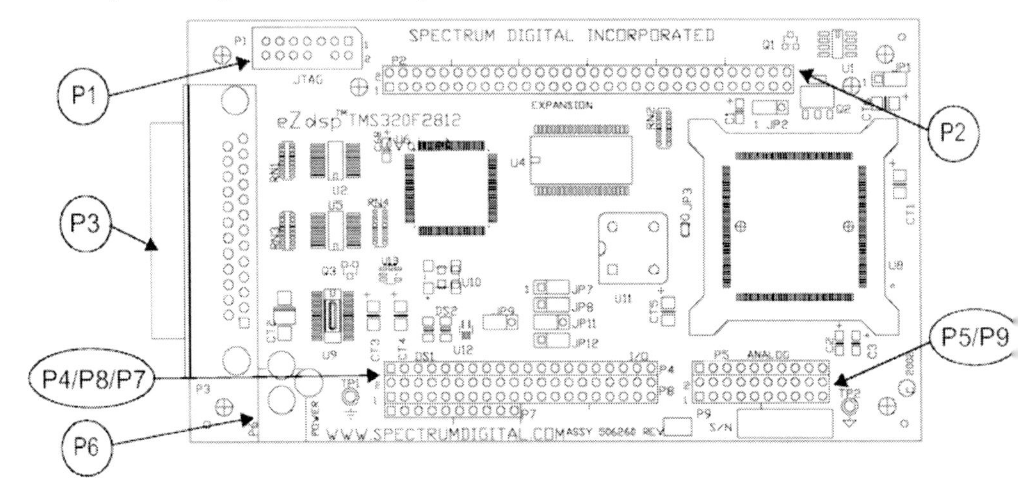

Figure 3.2 eZdsp F2812 Connector Positions [22]

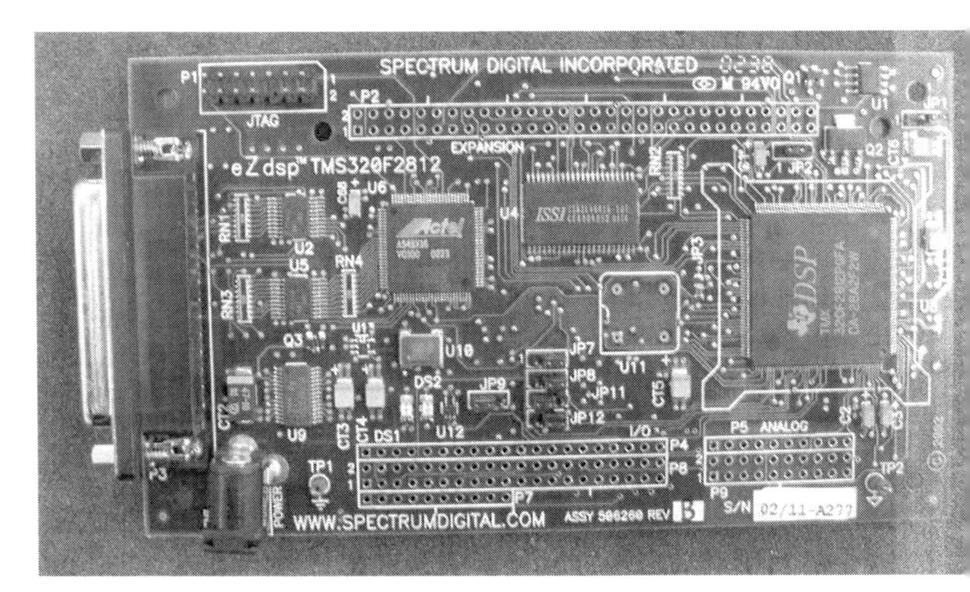

Figure 3.3 Top View of the eZdsp F2812 Board

The function of each connector is shown in the table below:

Connector	Function
P1	JTAG Interface
P2	Expansion
P3	Parallel Port/JTAG Controller Interface
P4/P7/P8	I/O Interface
P5/P9	Analog Interface
P6	Power Connector

Table 3.1 eZdsp F2812 Connectors

P1, JTAG Interface

The eZdsp F2812 is supplied with a 14-pin header interface, P1. This is the standard interface used by JTAG emulators to interface to Texas Instruments DSPs. The positions of the 14 pins on the P1 connector are shown in the diagram below as viewed from the top of the eZdsp.

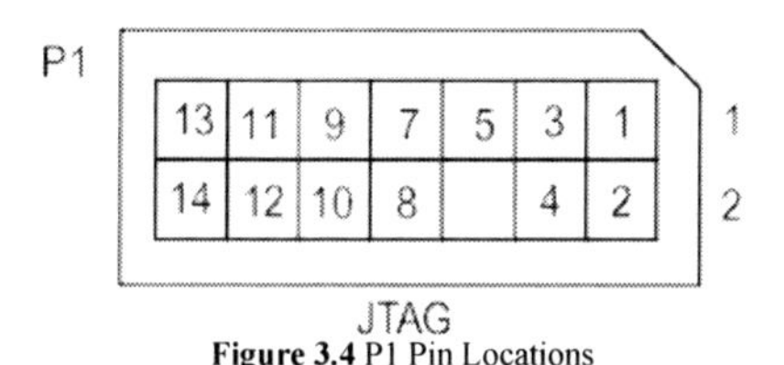

Figure 3.4 P1 Pin Locations

P2, Expansion Interface

The positions of the 60 pins on the P2 connector are shown in the diagram below as viewed from the top of the eZdsp.

P2

2	4	6	8	10	12	14	16	18	20	22	24	26	28	30	32	34	36	38	40	41	43	45	47	49	51	54	56	58	60
1	3	5	7	9	11	13	15	17	19	21	23	25	27	29	31	33	35	37	39	42	44	46	48	50	52	53	55	57	59

Figure 3.5 Connector P2 Pin Locations

P3, Parallel Port/JTAG Interface

The eZdsp F2812 uses a custom parallel port-JTAG interface device. The device has direct access to the integrated JTAG interface. Drivers for C2000 Code Composer tools are shipped with the eZdsp modules.

P4/P8/P7, I/O Interface

The connectors P4, P8, and P7 present the I/O signals from the DSP. The layouts of these connectors are shown below.

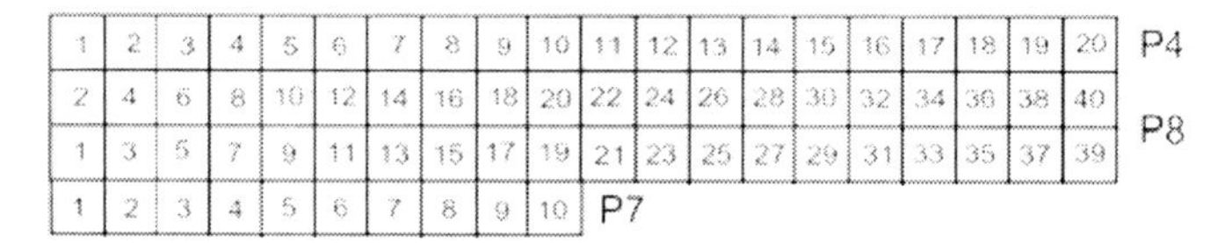

Figure 3.6 P4/P8/P7 Connectors

The pin definitions of P4/P8 connectors are shown in the table on the next page.

P8 Pin#	P8 Signal	P8 Pin#	P8 Signal
1	+5 Volts	2	+5 Volts
3	SCITXDA	4	SCIRXDA
5	XINT1n/XBIOn	6	CAP1/QEP1
7	CAP2/QEP2	8	CAP3/QEP3
9	PWM1	10	PWM2
11	PWM2	12	PWM4
13	PWM3	14	PWM6
15	T1PWM/T1CMP	16	T2PWM/T2CMP
17	TDIRA	18	TCLKINA
19	GND	20	GND
21	No Connect	22	XINT1N/XBIOn
23	SPISIMOA	24	SPISOMIA
25	SPICLKA	26	SPISTEA
27	CANTXA	28	CANRXA
29	XCLKOUT	30	PWM7
31	PWM8	32	PWM9
33	PWM10	34	PWM11
35	PWM12	36	CAP4/QEP3
37	T1CTRIP/PDPINTAn	38	T3CTRIP/PDPINTBn
39	GND	40	GND

Table 3.2 P8, I/O Connectors

P5/P9, Analog Interface

The positions of the 30 pins on the P5/P9 connectors are shown in the diagram below as viewed from the top of the eZdsp.

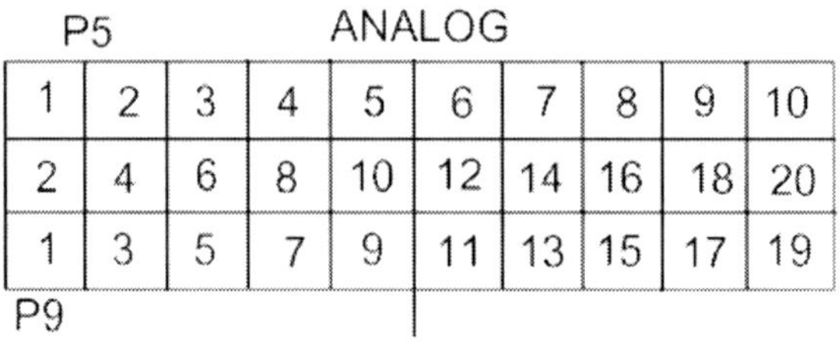

P5 ANALOG

1	2	3	4	5	6	7	8	9	10
2	4	6	8	10	12	14	16	18	20
1	3	5	7	9	11	13	15	17	19

P9

Figure 3.7 Connector P5/P9 Pin Locations

The definitions of P9 signals are shown in the table below.

P9 Pin#	Signal	P9 Pin#	Signal
1	GND	2	ADCINA0
3	GND	4	ADCINA1
5	GND	6	ADCINA2
7	GND	8	ADCINA3
9	GND	10	ADCINA4
11	GND	12	ADCINA5
13	GND	14	ADCINA6
15	GND	16	ADCINA7
17	GND	18	VREFLO
19	GND	20	No Connect

Table 3.3 P9, Analog Interface Connector

P6, Power Connector

Power (5 volts) is brought onto the eZdsp F2812 via the P6 connector

4. DSP Software Development

4.1 Basic Software Tools Required

In order to get started with DSP software development a number of basic tools will be needed. These will allow software routines to be written, converted to a form understandable to the DSP, and then downloaded to the target device. These tools were supplied with the development system as part of the complete package (CCS). The following list describes the basic tools required.

- **Basic text editor** This can be a very simple application such as DOS edit. It is used for entering DSP assembly language programs into a text file.
- **Assembler** It is used to convert the text based assembly program into a machine readable format (COFF object Files).
- **Linker** Organizes the machine readable code generated by the assembler so that it will match the memory configuration of the target DSP, i.e. DSP machine code.
- **Debug environment** Enables software to be tested for the particular DSP device; the debug environment may be in the form of a simulator or an emulator.
- **Downloader** Transfers the assembled and linked programs to the DSP development board.
- **Hex conversion utility** Converts assembled and linked DSP code into a form suitable for an EPROM programmer.

An overview of the steps taken during DSP software development is shown in Figure 4.1.

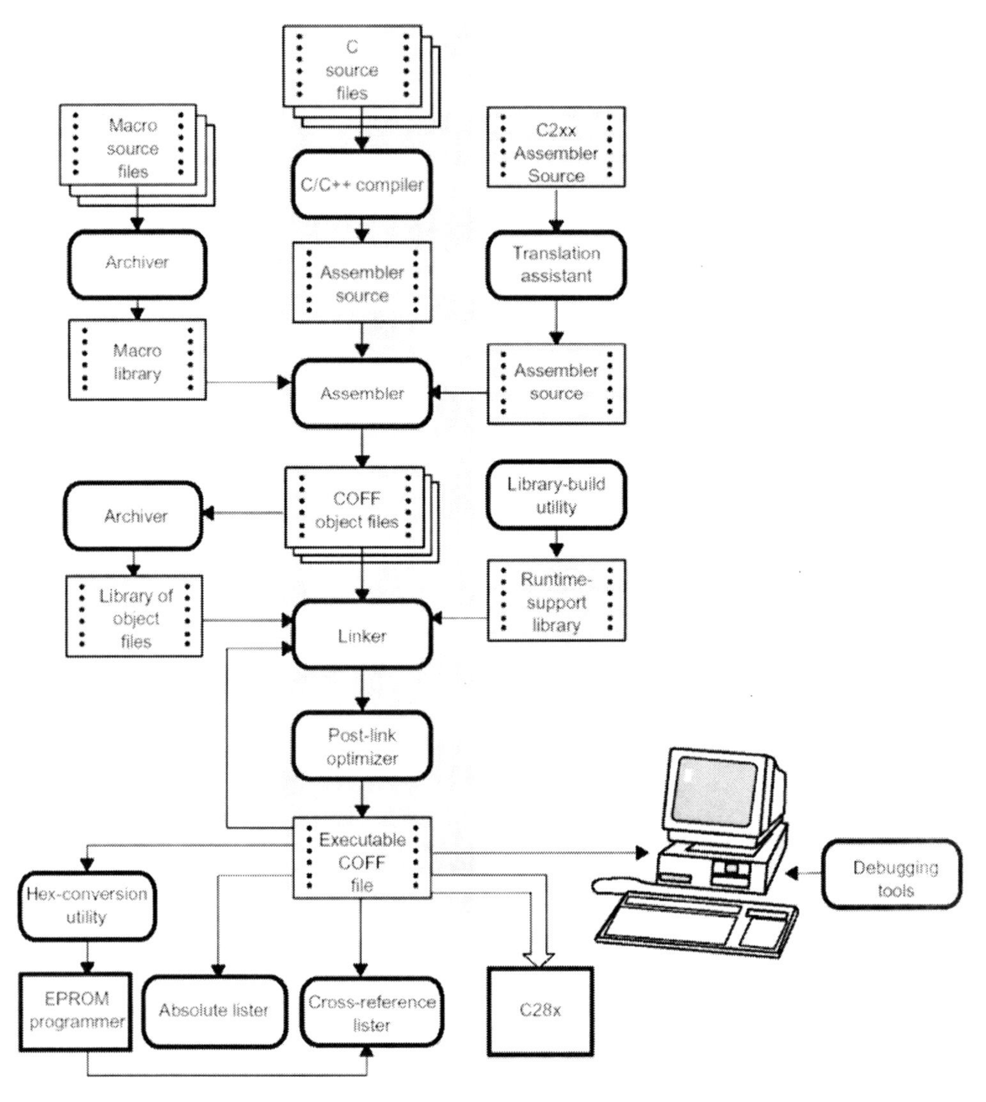

Figure 4.1 Steps Taken during DSP Software Development

4.2 Code Composer Studio

Designed for the Texas Instruments TMS320Cx digital signal processor (DSP) platforms, Code Composer Studio is a development environment that tightly integrates the capabilities of the following components within an extendible plug-in architecture:

- Integrated development environment (Code Composer) with editor, debugger, project manager, profiler, Probe Points, and more
- C Compiler, Assembly Optimizer and Linker (Code Generation Tools)
- Instruction Set Simulator
- Real-Time Foundational Software (DSP/BIOS)
- Real-Time Data Exchange Between Host and Target (RTDX)
- Real-Time Analysis and Data Visualization

4.2.1 Creating a New Project

The following procedure allows to create new projects, either individually or several at once. Each project's filename must be unique. The information for a project is stored in a single project file (*.pjt).

Step 1: From the Project menu, choose New. The Project Creation wizard displays:

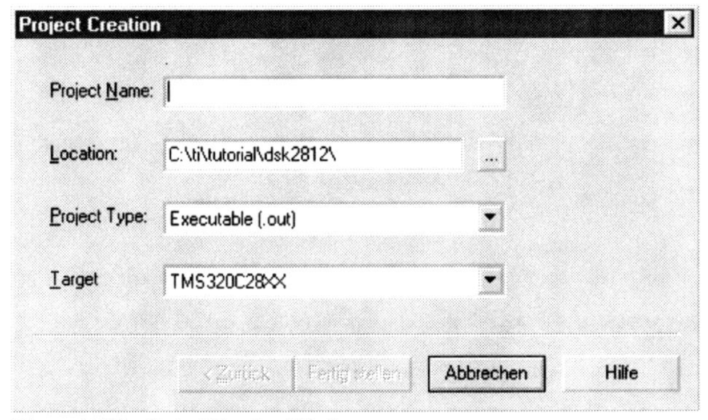

Figure 4.2 The Project Creation Window

Step 2: In the Project Name field, type a name for the new project. Each project must have a unique name.

Step 3: In the Location field, specify a directory to store the project file. It is a good idea to use a different directory for each new project. Use this directory to store project files and the object files generated by the compiler and assembler.

Step 4: In the Project Type field, select a Project Type from the drop-down list. Choose either Executable (.out) or Library (lib). Executable indicates that the project generates an executable file. Library indicates that you are building an object library.

Step 5: In the Target field, select the Target Family that identifies your CPU. This information is necessary when tools are installed for multiple targets.

Step 6: Click Finish.

The CCStudio IDE creates a project file called projectname.pjt. This file stores the project settings and references the various files used by the project. The new project automatically becomes the active project. The first project configuration (in alphabetical order) is set as active. The new project inherits TI-supplied default compiler and linker options for debug and release configurations.

After creating a new project file, add the filenames of the source code, object libraries, and linker command file to the project list.

4.2.2 Adding Files to a Project

You can add several different files or file types to a project. Here is the procedure to add files to a project:

Step 1: Select Project→Add Files to Project, or right-click on the project's filename in the Project View window and select Add Files. The Add Files to Project dialog box displays.

Step 2: In the Add Files to Project dialog box, specify a file to add. Use the Files of type drop-down list to set the type of files that appear in the File name field.

Note: Do not try to manually add header/include files (*.h) to the project. These files are automatically added when the source files are scanned for dependencies as part of the build process.

Step 3: Click Open to add the specified file to the project. The Project View is automatically updated when a file is added to the current project.

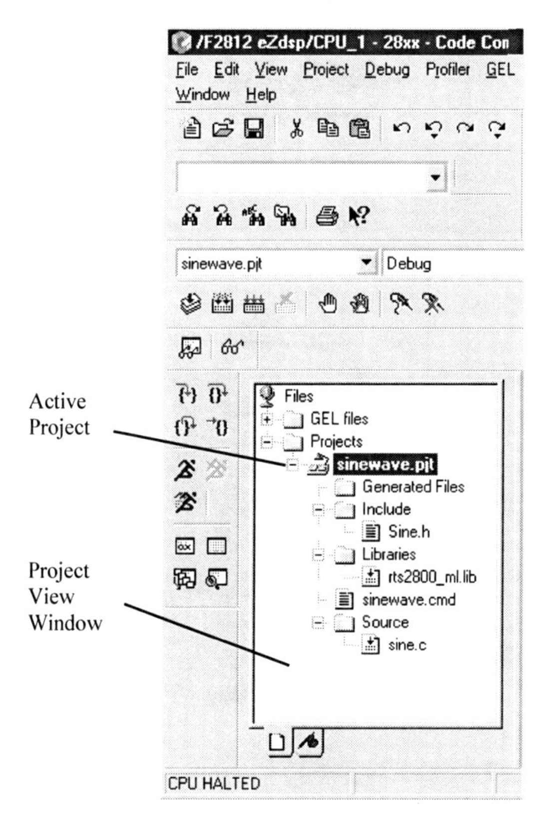

Figure 4.3 The Project Manager Window

The project manager organizes files into folders for source files, include files, libraries, and DSP/BIOS configuration files. Source files that are generated by DSP/BIOS are placed in the Generated files folder.

4.2.3 Building and Running the Program

To build and run a program, the following steps must be carried out:

Step 1: Choose Project→Rebuild All or click the (Rebuild All) toolbar button. The CCStudio IDE recompiles, reassembles, and relinks all the files in the project. Messages about this process are shown in a frame at the bottom of the window.

Step 2: By default, the .out file is built into a debug directory located under the current project folder. To change this location, select a different one from the CCStudio toolbar.

Step 3: Choose File→Load Program. Select the program you just rebuilt, and click Open. The CCStudio IDE loads the program onto the target DSP and opens a Dis-Assembly window that shows the disassembled instructions that make up the program.

Step 4: Choose View→Mixed Source/ASM. This allows simultaneous view of the c source and the resulting assembly code.

Step 5: Choose Debug→Go Main to begin execution from the main function.

Step 6: Choose Debug→Run or click the (Run) toolbar button to run the program.

Step 7: Choose Debug→Halt to quit running the program.

4.2.4 Introduction to Breakpoints

Breakpoints are an essential component of any debugging session. Breakpoints stop the execution of the program. While the program is stopped, you can examine the state of the program, examine or modify variables, examine the call stack, etc. Breakpoints can be set on a line of source code in an Editor window or a disassembled instruction in the Disassembly window. After a breakpoint is set, it can be enabled or disabled.

For breakpoints set on source lines it is necessary that there be an associated line of disassembly code. When compiler optimization is turned on, many source lines cannot have breakpoints set. To see allowable lines, use mixed mode in the editor window.

Breakpoints can be set in any Disassembly window or document window containing C/C++ source code. There is no limit to the number of software breakpoints that can be set, provided they are set at writable memory locations (RAM). Software breakpoints operate by modifying the target program to add a breakpoint instruction at the desired location.

The fastest way to set a breakpoint is to simply double-click on the desired line of code.

Step 1: In a document window or Disassembly window, move the cursor over the line where you want to place a breakpoint.

Step 2: Double-click in the Selection Margin immediately preceding the line when you are in a document window.

In a Disassembly window, double-click on the desired line.

A breakpoint icon in the Selection Margin indicates that a breakpoint has been set at the desired location.

The Toggle Breakpoint command and the Toggle Breakpoint button also enable you to quickly set and clear breakpoints.

Step 1: In a document window or Disassembly window, put the cursor in the line where you want to set the breakpoint.

Step 2: Right-click and select Toggle Breakpoint, or click the Toggle Breakpoint button on the Project toolbar.

4.2.5 Watch Window

When debugging a program, it is often helpful to understand how the value of a variable changes during program execution. The Watch window allows monitoring the values of local and global variables and C/C++ expressions. To open the Watch window: Select View→Watch Window, or click the Watch Window button on the Watch toolbar.

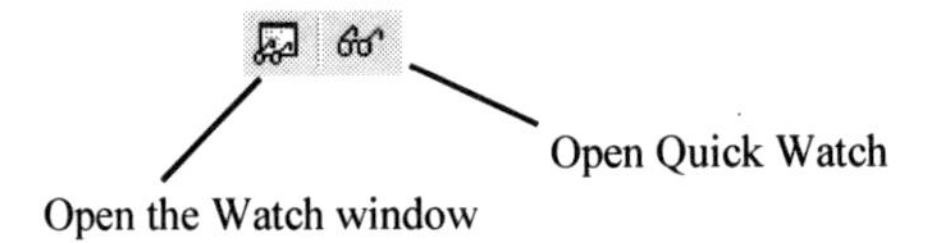

The Watch window, Figure 4.4, contains two tabs labeled: Watch Locals and Watch.

- In the Watch Locals tab, the debugger automatically displays the Name, Value, and Type of the variables that are local to the currently executing function.
- In the Watch tab, the debugger displays the Name, Value, and Type of the local and global variables and expressions that you specify.

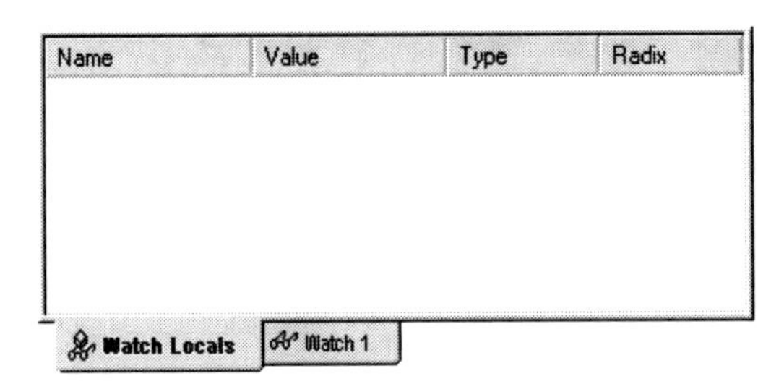

Figure 4.4 The Watch Window

4.2.6 Probe Points

A Probe Point reads data from a file on PC. Probe Points are a useful tool for algorithm development. Probe points can be used to:

- Transfer input data from a file on the host PC to a buffer on the target for use by the algorithm.
- Transfer output data from a buffer on the target to a file on the host PC for analysis.
- Update a window, such as a graph, with data.

Probe Points are similar to breakpoints in that they both halt the target to perform their action. However, Probe Points differ from breakpoints in the following ways:

- Probe Points halt the target momentarily, perform a single action, and resume target execution.
- Breakpoints halt the CPU until execution is manually resumed and cause all open windows to be updated.
- Probe Points permit automatic file input or output to be performed; breakpoints do not.

This section shows how to use a Probe Point to transfer the contents of a PC file to the target for use as test data. It also uses a breakpoint to update all the open windows when the Probe Point is reached.

Step 1: Choose File. Load Program. Select *filename*.out, and click Open.

Step 2: Double-click on the *filename*.c file in the Project View.

Step 3: Put the cursor in a line of the main function to which you want to add a probe point.

Step 4: Click the (Toggle Probe Point) toolbar button.

Step 5: From the File menu, choose File I/O. The File I/O dialog appears so that you can select input and output files.

Step 6: In the File Input tab, click Add File.

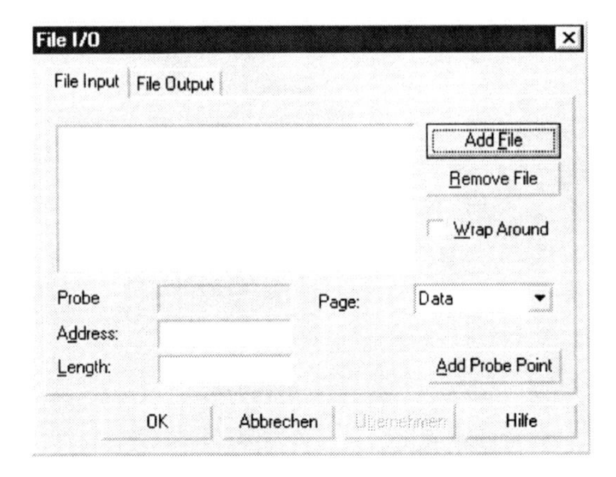

Figure 4.5 The File I/O Dialog

Step 7: Browse to the project folder, select *filename*.dat and click Open. A control window for the *filename*.dat file appears. When you run the program, you can use this window to start, stop, rewind, or fast forward within the data file:

Step 8: In the File I/O dialog, change the Address and the Length values. Also, put a check mark in the Wrap Around box:

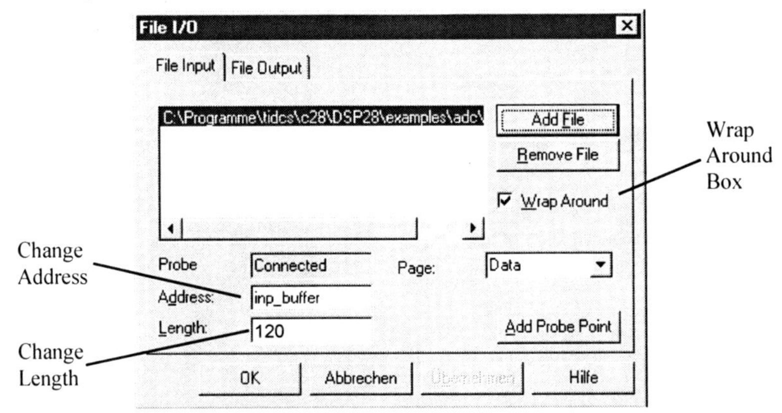

Figure 4.6 Changing the Address and Length Values in the File I/O Dialog

The Address field specifies where the data from the file is to be placed.

- The Length field specifies how many samples from the data file are read each time the Probe Point is reached.
- The Wrap Around option causes the CCStudio IDE to start reading from the beginning of the file when it reaches the end of the file. This allows the data file to be treated as a continuous stream of data.

Step 9: Click Add Probe Point. The Probe Points tab of the Break/Probe Points dialog appears.

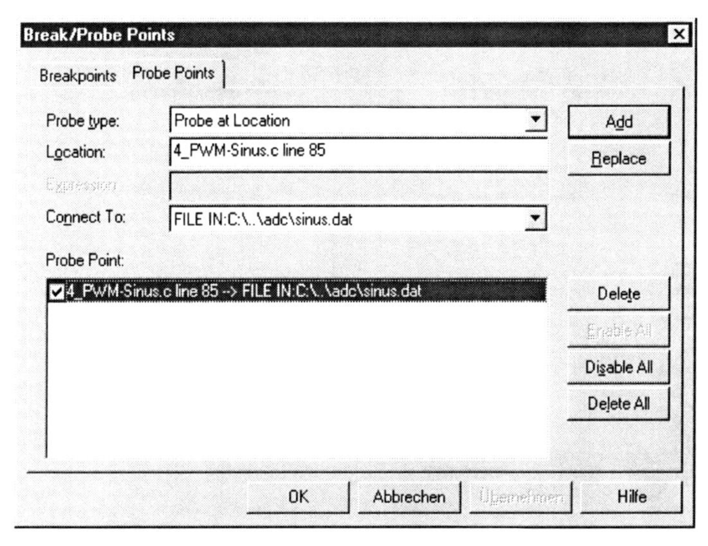

Figure 4.7 The Brake/Probe Points Dialog

Step 10: In the Probe Point list, highlight a line.

Step 11: In the Connect To field, click the down arrow and select a file from the list.

Step 12: Click Replace. The Probe Point list changes to show that this Probe Point is connected to the sinus.dat file.

Step 13: Click OK. The File I/O dialog shows that the file is now connected to a Probe Point.

4.2.7 Displaying Graphs

If you run a program using only breakpoints and watch windows, you would not see much information about what the program was doing. You could set watch variables on addresses within the inp_buffer and out_buffer arrays, but you would need to watch a lot of variables and the display would be numeric rather than visual.

CCStudio IDE provides a variety of ways to graph data processed by your program. In this example, you view a signal plotted against time.

Step 1: Choose View→Graph→Time/Frequency. The Graph Property Dialog box appears.

Step 2: In the Graph Property Dialog, change the Graph Title, Start Address, Acquisition Buffer Size, Display Data Size, DSP Data Type, Autoscale, and Maximum Y-value properties according to your program.

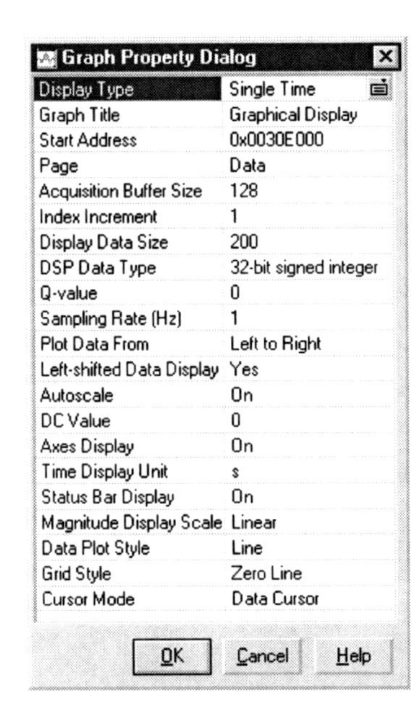

Figure 4.8 The Graph Property Dialog

Step 3: Click OK. A graph window for the Input Buffer appears.

Step 4: Right-click on the Input window and choose Clear Display from the pop-up menu.

Step 5: Choose View→Graph→Time/Frequency again.

Step 6: This time, change the Graph Title to Output and the Start Address to out_buffer. All the other settings are correct.

Step 7: Click OK to display the graph window for the Output. Right-click on the graph window and choose Clear Display from the pop-up menu.

5. PWM

5.1 Definition

PWM means Pulse Width Modulation, one variation of which is keeping the frequency constant and changing the duty-cycle of the waveform. A PWM signal is a sequence of pulses with changing pulse widths. The pulses are spread over a number of fixed-length periods so that there is one pulse in each period. The fixed period is called the PWM (carrier) period and its inverse is called the PWM (carrier) frequency. The widths of the PWM pulses are determined, or modulated, from pulse to pulse according to another sequence of desired values, the modulating signal. Typical PWM applications include motor speed control, battery chargers, and switching voltage regulators.

In a motor control system, PWM signals are used to control the on and off time of switching power devices that deliver the desired current and energy to the motor windings. The shape and frequency of the phase currents and the amount of energy delivered to the motor windings control the required speed and torque of the motor. In this case, the command voltage or current to be applied to the motor is the modulating signal. The frequency of the modulating signal is typically much lower than the PWM carrier frequency [21].

The pulse width modulated (PWM) signal outputs on a TMS320C2812 DSP are variable duty cycle square-waves with 5 volt amplitude. These signals can each be decomposed into a D.C. component plus a new square-wave of identical duty-cycle but with a time-average amplitude of zero. Figure 5.1 depicts this graphically [23]. It is obvious, that the amplitude of the D.C. component is directly proportional to the PWM duty cycle.

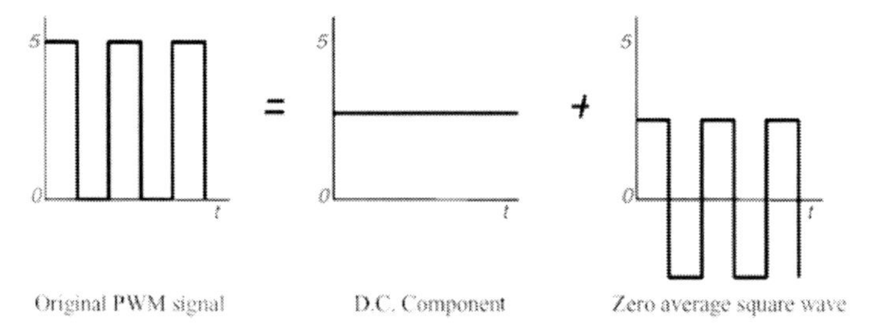

Figure 5.1 Decomposition of PWM Signal (shown for 50% duty cycle)

5.2 Event Manager PWM Waveform Generation

Up to eight PWM waveforms (outputs) can be generated simultaneously by each event manager: three independent pairs (six outputs) by the three full-compare units with programmable deadbands, and two independent PWMs by the GP-timer compares.

PWM Characteristics

Characteristics of the PWMs are as follows:

- 16-bit registers
- Wide range of programmable deadband for the PWM output pairs
- Change of the PWM carrier frequency as needed
- Change of the PWM pulse widths within and after each PWM period as needed
- Pulse-pattern-generator circuit, for programmable generation of asymmetric, symmetric, and four-space vector PWM waveforms

To generate a PWM signal, an appropriate timer is needed to repeat a counting period that is the same as the PWM period. A compare register is used to hold the modulating values. The value of the compare register is constantly compared with the value of the timer counter. When the values match, a transition (from low to high, or high to low) happens on the associated output. When a second match is made between the values, or when the end of a timer period is reached, another transition (from high to low or low to high) happens on the associated output. In this way, an output pulse is generated whose on (or off) duration is proportional to the value in the compare register. This process is repeated for each timer period with different (modulating) values in the compare register. As a result, a PWM signal is generated at the associated output.

Dead Band

In many motion/motor and power electronics applications, two power devices, an upper and a lower, are placed in series on one power converter leg. The turn-on periods of the two devices must not overlap with each other in order to avoid a shoot-through fault. Thus, a pair of non-overlapping PWM outputs is often required to properly turn on and off the two devices. A dead time (deadband) is often inserted between the turning-off of one transistor and the turning-on of the other transistor. This delay allows complete turning-off of one transistor before the turning-on of the other transistor. The required time delay is specified by the turning-on and turning-off characteristics of the power transistors and the load characteristics in a specific application.

5.3 Generation of PWM Outputs

Each of the three compare units, together with GP timer 1 (in the case of EVA) or GP timer 3 (in the case of EVB), the dead-band unit, and the output logic in the event manager module, can be used to generate a pair of PWM outputs with programmable dead-band and output polarity on two dedicated device pins. There are six such dedicated PWM output pins associated with the three compare units in each EV module. These six dedicated output pins can be used to conveniently control 3-phase AC induction or brushless DC motors. The flexibility of output behavior control by the compare action control register (ACTRx) also makes it easy to control switched reluctance and synchronous reluctance motors in a wide range of applications. The PWM circuits can also be used to conveniently control other types of motors such as DC brush and stepper motors in single or multi-axis control applications. Each GP timer compare unit, if desired, can also generate a PWM output based on its own timer [21].

5.3.1 Asymmetric and Symmetric PWM Generation

Both asymmetric and symmetric PWM waveforms can be generated by every compare unit on the EV module. All three kinds of PWM waveform generations with compare units and associated circuits require configuration of the same Event Manager registers. The setup process for PWM generation includes the following steps:

- Setup and load ACTRx
- Setup and load DBTCONx, if dead-band is to be used
- Initialize CMPRx
- Setup and load COMCONx
- Setup and load T1CON (for EVA) or T3CON (for EVB) to start the operation
- Rewrite CMPRx with newly determined values

Asymmetric PWM Waveform Generation

The edge-triggered or asymmetric PWM signal is characterized by modulated pulses, which are not centered with respect to the PWM period, as shown in Figure 5.2 [21]. The width of each pulse can only be changed from one side of the pulse. Such software controlled flexibility of PWM outputs is particularly useful in switched reluctance motor control applications.

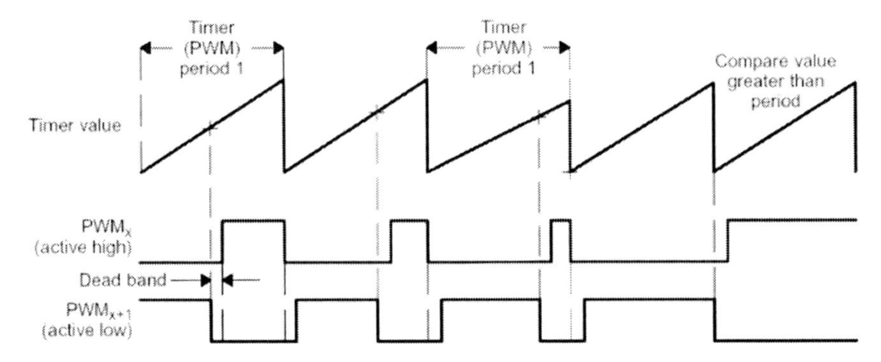

Figure 5.2 Asymmetric PWM Waveform Generation with Compare Unit and PWM Circuits (x = 1, 3, or 5)

Symmetric PWM Waveform Generation

A centered or symmetric PWM signal is characterized by modulated pulses, which are centered with respect to each PWM period. The advantage of a symmetric PWM signal over an asymmetric PWM signal is that it has two inactive zones of the same duration: at the beginning and at the end of each PWM period. This symmetry has been shown to cause less harmonics than an asymmetric PWM signal in the phase currents of an AC motor, such as induction and DC brushless motors, when sinusoidal modulation is used. Figure 5.3 shows two examples of symmetric PWM waveforms.

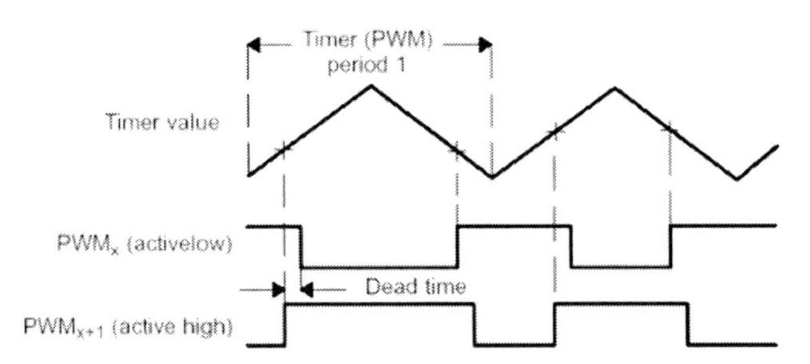

Figure 5.3 Symmetric PWM Waveform Generation with Compare Units and PWM Circuits (x = 1, 3, or 5) [21]

5.3.2 Program Example

To describe the generating of PWM waveforms using the Event Manager Module, here an example program code is explained. This program sets up the EV timers (TIMER1, TIMER2, TIMER3 and TIMER4) to generate T1PWM, T2PWM, T3PWM, T4PWM and PWM1-12 waveforms. The waveforms can then be observed using a scope. This program requires the DSP28 header files. To compile the program as it is the header files should be added to the program in the CCS.

```
// Step 1. Include required header files
        // DSP28_Device.h: device specific definitions

#include "DSP28_Device.h"

void main(void)
{
// Step 2. Initialize System Control registers, PLL, WatchDog, Clocks to default state:
    // This function is found in the DSP28_SysCtrl.c file.
        InitSysCtrl();

// Step 3. Select GPIO for the device or for the specific application:
    // This function is found in the DSP28_Gpio.c file.
        // Initialize GPIO for this test here
        EALLOW;
        // Enable PWM pins
    GpioMuxRegs.GPAMUX.all = 0x00FF; // EVA PWM 1-6  pins
    GpioMuxRegs.GPBMUX.all = 0x00FF; // EVB PWM 7-12 pins
    EDIS; // Disable PWM pins

// Step 4. Initialize PIE vector table:
        // The PIE vector table is initialized with pointers to shell Interrupt
    // Service Routines (ISR).  The shell routines are found in DSP28_DefaultIsr.c.
        // Insert user specific ISR code in the appropriate shell ISR routine in
    // the DSP28_DefaultIsr.c file.
```

```
// Disable and clear all CPU interrupts:
DINT;
IER = 0x0000;
IFR = 0x0000;

// Initialize Pie Control Registers To Default State:
// This function is found in the DSP28_PieCtrl.c file.
InitPieCtrl();

// Initialize the PIE Vector Table To a Known State:
// This function is found in DSP28_PieVect.c.
// This function populates the PIE vector table with pointers
// to the shell ISR functions found in DSP28_DefaultIsr.c.
InitPieVectTable();

// EVA Configure T1PWM, T2PWM, PWM1-PWM6
// Step 1 Initialize the timers
// Initialize EVA Timer1
EvaRegs.T1PR = 0xFFFF;       // Timer1 period
EvaRegs.T1CMPR = 0x3C00;     // Timer1 compare
EvaRegs.T1CNT = 0x0000;      // Timer1 counter
// TMODE = continuous up/down
// Timer enable
// Timer compare enable
EvaRegs.T1CON.all = 0x1042;

// Initialize EVA Timer2
EvaRegs.T2PR = 0x0FFF;       // Timer2 period
EvaRegs.T2CMPR = 0x03C0;     // Timer2 compare
EvaRegs.T2CNT = 0x0000;      // Timer2 counter
```

```
   // TMODE = continuous up/down
        // Timer enable
        // Timer compare enable
        EvaRegs.T2CON.all = 0x1042;

// Step 2  Setup T1PWM and T2PWM
        // Drive T1/T2 PWM by compare logic
        EvaRegs.GPTCONA.bit.TCOMPOE = 1;
        // Polarity of GP Timer 1 Compare = Active low
        EvaRegs.GPTCONA.bit.T1PIN = 1;
        // Polarity of GP Timer 2 Compare = Active high
        EvaRegs.GPTCONA.bit.T2PIN = 2;

// Step 3 Enable compare for PWM1-PWM6
        EvaRegs.CMPR1 = 0x0C00;
        EvaRegs.CMPR2 = 0x3C00;
        EvaRegs.CMPR3 = 0xFC00;

   // Compare action control.  Action that takes place on a compare event
   // output pin 1 CMPR1 - active high
   // output pin 2 CMPR1 - active low
   // output pin 3 CMPR2 - active high
   // output pin 4 CMPR2 - active low
   // output pin 5 CMPR3 - active high
   // output pin 6 CMPR3 - active low
   EvaRegs.ACTRA.all = 0x0666;
        EvaRegs.DBTCONA.all = 0x0000; // Disable deadband
   EvaRegs.COMCONA.all = 0xA600;

// EVB Configure T3PWM, T4PWM and PWM7-PWM12
// Step 1 - Initialize the Timers
```

```
// Initialize EVB Timer3
        // Timer3 controls T3PWM and PWM7-12
        EvbRegs.T3PR = 0xFFFF;        // Timer3 period
        EvbRegs.T3CMPR = 0x3C00;      // Timer3 compare
        EvbRegs.T3CNT = 0x0000;       // Timer3 counter
// TMODE = continuous up/down
        // Timer enable
        // Timer compare enable
        EvbRegs.T3CON.all = 0x1042;

// Initialize EVB Timer4
// Timer4 controls T4PWM
        EvbRegs.T4PR = 0x00FF;        // Timer4 period
        EvbRegs.T4CMPR = 0x0030;      // Timer4 compare
        EvbRegs.T4CNT = 0x0000;       // Timer4 counter

// TMODE = continuous up/down
        // Timer enable
        // Timer compare enable
        EvbRegs.T4CON.all = 0x1042;

// Step 2  Setup T3PWM and T4PWM
        // Drive T3/T4 PWM by compare logic
        EvbRegs.GPTCONB.bit.TCOMPOE = 1;
 // Polarity of GP Timer 3 Compare = Active low
        EvbRegs.GPTCONB.bit.T3PIN = 1;
        // Polarity of GP Timer 4 Compare = Active high
        EvbRegs.GPTCONB.bit.T4PIN = 2;

// Step 3  Enable compare for PWM7-PWM12
        EvbRegs.CMPR4 = 0x0C00;
        EvbRegs.CMPR5 = 0x3C00;
        EvbRegs.CMPR6 = 0xFC00;
```

```
    // Compare action control.  Action that takes place on a compare event
    // output pin 1 CMPR4 - active high
    // output pin 2 CMPR4 - active low
    // output pin 3 CMPR5 - active high
    // output pin 4 CMPR5 - active low
    // output pin 5 CMPR6 - active high
    // output pin 6 CMPR6 - active low
    EvbRegs.ACTRB.all = 0x0666;
        EvbRegs.DBTCONB.all = 0x0000; // Disable deadband
    EvbRegs.COMCONB.all = 0xA600;

// Step 6. IDLE loop. Loop forever:
//  PWM pins can be observed with a scope.
        for(;;);
}
```

As a result of this program, one can observe the PWM signals and their complemented pairs with an oscilloscope at the PWM pins. Some examples are given in the figures below:

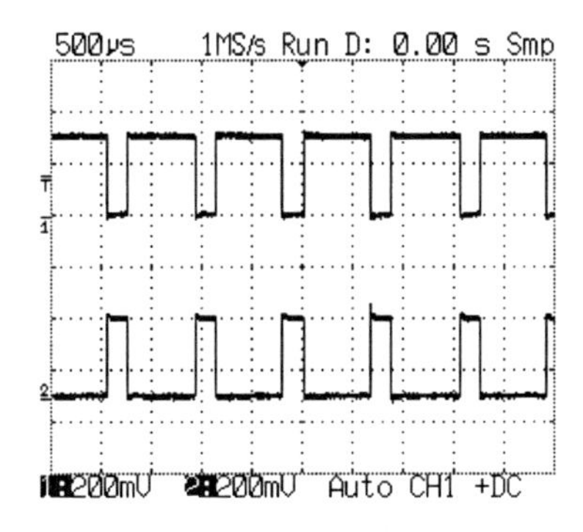

Figure 5.4 PWM Signal and Its Complement at the Pins PWM3&4 and PWM9&10

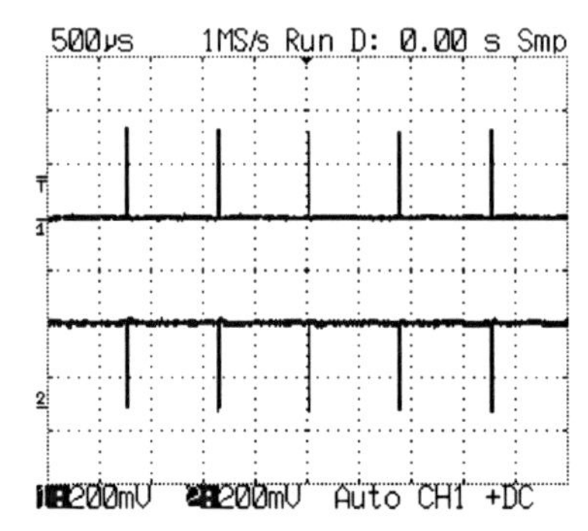

Figure 5.5 PWM Signal and Its Complement at the Pins PWM5&6 and PWM11&12

5.3.3 Dead-Time Generation on the TMS320C2812

A Dead-Band generator protects the power semiconductors during the commutation. The dead time is programmable between 0μs and 448μs. This time range is sufficient for all kinds of power semi-conductors (MOSFET, IGBT, BIPOLAR, and THYRISTOR) in a wide range of kW (kilowatts) or HP (Horse Powers).

TMS320C2812 Dead-Band separates the transition edges of two signals: output and complemented output, by a time interval. This time interval is programmable. The Dead-Band can only be used with Full-compare Output.

Full-compare has two outputs per channel, a "true" phase and a "false" phase. These exits allow the device to directly drive the upper and the lower halves of an H-bridge. To accommodate any combinations of transistor types and polarities in the H-bridge, the state of the outputs in the ACTIVE versus INACTIVE time slots are programmable. Therefore, it does not necessarily follow that, in the dual output compare channels, the true and false phase outputs are electrically complementary. Furthermore, it is also not true that when one output is ACTIVE, the other is INACTIVE.

The key point to remember is:

- Both "true" and "false" phase outputs use the same definition for ACTIVE vs. INACTIVE time slots. The electrical state of the output pins will be determined by the value programmed in the appropriate ACTION register (ACTR) for the ACTIVE state.

In fact, the only distinguished feature between the true and false outputs is the generation of the dead band time. Dead-Band generation is accomplished by digitally counting a programmable number of cycles between the generate edges on the true and false outputs due to the compare trigger event. The delay generated starts when the compare event happens.

The rules for Dead-Band generation are:

- When a compare event happens to enter the ACTIVE time slot of the PWMcycle, the FALSE output changes from the INACTIVE state to the ACTIVE state immediately. The TRUE output waits for the dead-band time before changing from its INACTIVE to ACTIVE state.
- When a compare or period occurs to enter the INACTIVE time slot of the PWM cycle, the TRUE side output changes from the ACTIVE state to the INACTIVE state immediately, while the FALSE side output changes after the dead-band time.
- If the time slot definition is reset to INACTIVE on an underflow event (symmetric only), both outputs go to the INACTIVE state immediately, no dead-band is generated [24].

Possibilities of Dead-Band Generator

The Dead-Band Generator associated with the General Purpose Timer One can be used:

- To generate three symmetrical PWM plus three complemented PWM with Dead-Time on the Full-compare output:

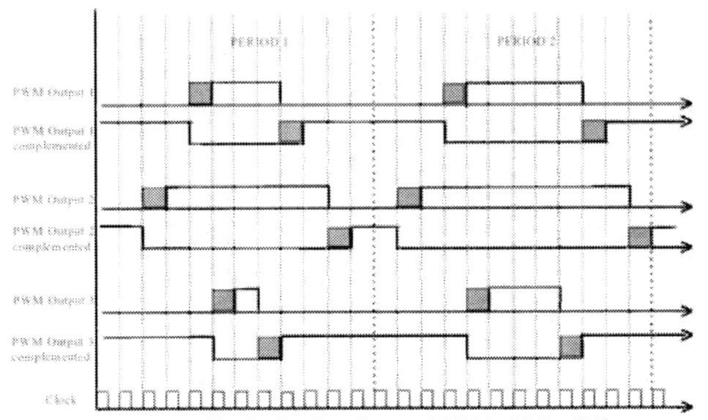

Figure 5.6 PWM Signals plus Complemented PWM with Dead-Time

- To generate three symmetrical PWM plus three PWM with Dead-time on the Full-compare output:

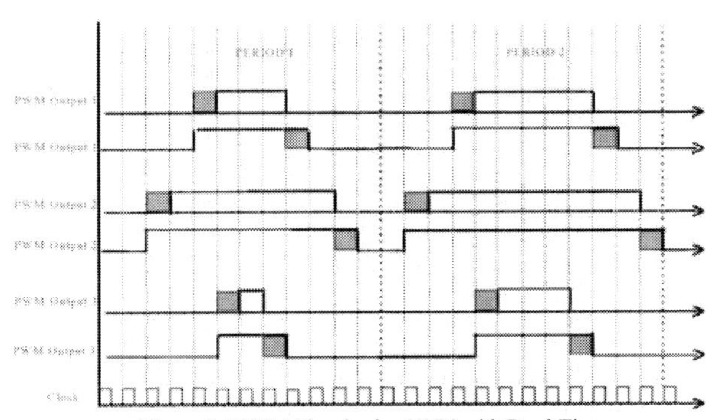

Figure 5.7 PWM Signals plus PWM with Dead-Time

5.3.3.1 Configuring PWM Outputs with Dead Band

The DSP controller TMS320F2812 has a programmable dead band generator, to insert a dead band between two PWM outputs (PWM1&2, PWM3&4, PWM5&6, PWM7&8, PWM9&10, and PWM11&12). The polarity of PWM channels (active high or active low) can be controlled using the ACTR register. The appropriate dead band between two PWM outputs can be inserted using the DBTCON register. The dead band circuitry of the TMS320F2812 reduces the active portion of the odd PWM channels from the leading transition end. However, the dead band circuitry increases the active portion of the even PWM channels from the lagging transition end. This part describes how to configure the PWM outputs with dead band for different power devices. The following figures show different PWM waveform configurations with dead band.

Figure 5.8 shows PWM1 (active high) and PWM2 (active low) without dead band. The counter is operating in continuous up/down mode and the transitions in PWM outputs occur at every compare match. The compare match event is shown by dotted lines in Figure 5.9 through Figure 5.12 [25].

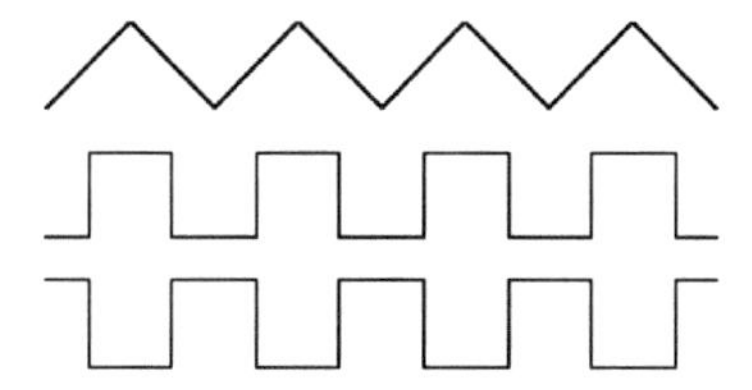

Figure 5.8 PWM1 & PWM2 without Dead Band: PWM1 (top) Active High and PWM2 (bottom) Active Low

Figure 5.9 shows PWM1 and PWM2 configured as active high and active low respectively, with dead band.

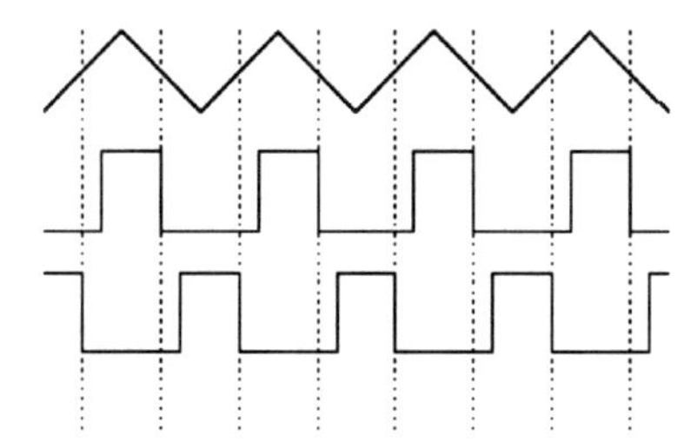

Figure 5.9 PWM1 (Active High) and PWM2 (Active Low) Outputs with Dead Band

Figure 5.10 shows PWM1 and PWM2 configured as active low and active high respectively, with dead band.

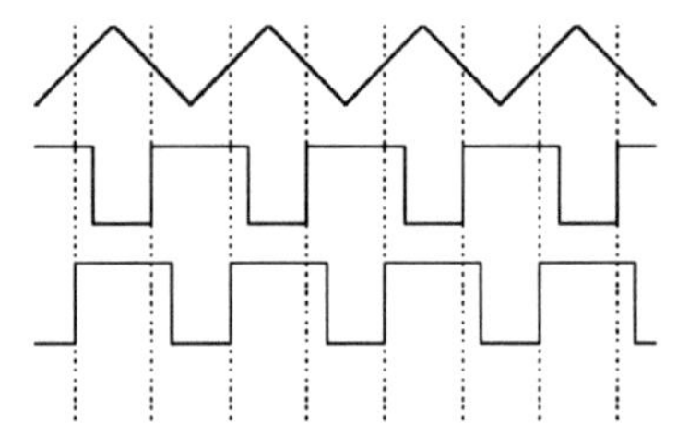

Figure 5.10 PWM1 (Active Low) and PWM2 (Active High) Outputs with Dead Band

Figure 5.11 shows both PWM1 and PWM2 configured as active high with dead band.

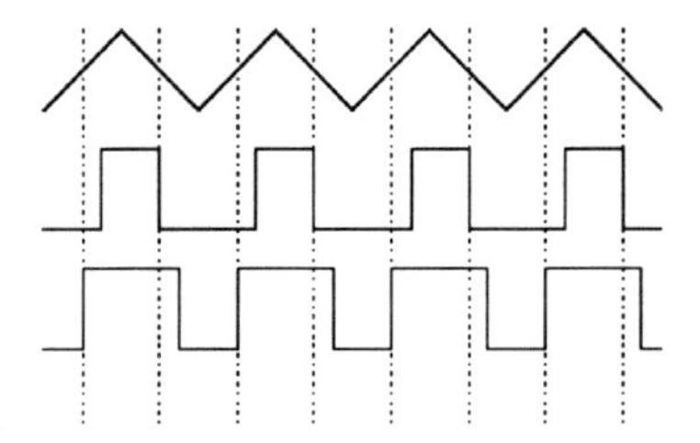

Figure 5.11 Both PWM1 and PWM2 Configured Active High with Dead Band

Figure 5.12 shows both PWM1 and PWM2 configured as active low with dead band.

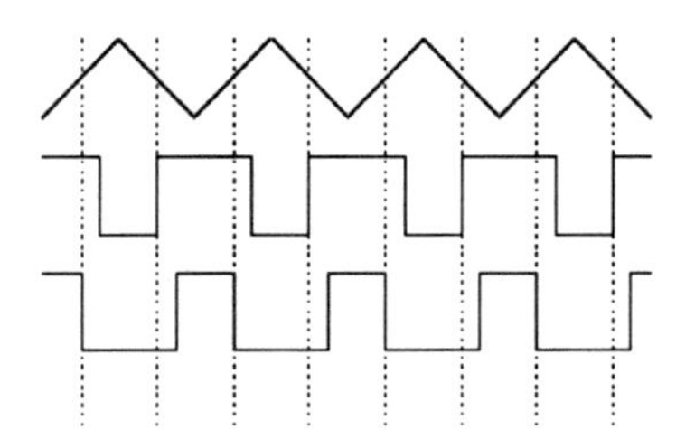

Figure 5.12 Both PWM1 and PWM2 Configured as Active Low with Dead Band

Note that in all the cases shown in Figure 5.9 through Figure 5.12, the active portion of PWM1 is reduced and the active portion of PWM2 is increased by the introduction of dead band.

Example

Figure 5.13 shows a phase of an inverter that needs dead band to prevent shoot through fault.

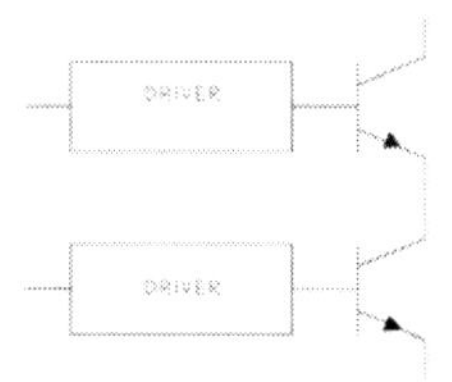

Figure 5.13 Inverter Phase with Two Power Devices Connected in Series

The dead-band unit is designed to prevent an overlap under any operating situation between the turn-on period of the upper and lower devices controlled by the two PWM outputs associated with each compare unit.

Suppose that both power devices turn on when the driver gets high voltage, and turn off when the driver gets low voltage. In this case the PWM outputs with dead band should be configured as shown in Figure 5.9.

If both power devices turn on with low voltage and turn off with high voltage, then the PWM outputs with dead band should be configured as shown in Figure 5.10.

If the upper device needs high voltage to turn on and the lower device needs low voltage to turn on, then the PWM outputs with dead band should be configured as shown in Figure 5.11.

If the upper device needs low voltage to turn on and the lower device needs high voltage to turn on, then the PWM outputs with dead band should be configured as shown in Figure 5.12.

5.4 Creating a PWM Signal with Fixed Duty Cycle and Frequency

This application generates a pulse width modulated (PWM) signal asymmetrically (edge-triggered) using C/C++ code. The PWM frequency is 30kHz with a 50% duty cycle.

To create an asymmetric PWM signal, the timer is set to the Continuous-Up Count Mode. (If a symmetric PWM signal is desired, then the Timer should be set to the Continuous-Up/Down Count Mode.) Furthermore, to create a consistent duty cycle for the PWM, the compare register should be loaded when the counter value is 0 or when the counter equals the value in the period register. Since the compare value is constant in this application, the selection of the reload condition for the compare register is not crucial. In other applications where the compare value will be changed, the reload condition of the compare register becomes important [26].

To create a PWM signal with a specific frequency and duty cycle, the values for the period and the compare registers need to be calculated. In order to calculate the values, the proper SYSCLKOUT frequency needs to be known; in this application, it is 150MHz.

Period register value calculation:

$$PeriodValue = \frac{HSPCLK}{2*(DesiredFrequency)} \quad (5\text{-}1)$$

where HSPCLK = SYSCLKOUT / (Input Clock Prescaler). (5-2)

The compare register value calculation:

Compare Value = Period Value x (Duty Cycle/100) (5-3)

In our case: Period Value = (150MHz)/(2*30kHz) = 2500. Accordingly, Compare Value = 2500*(50/100) = 1250.

To determine whether the polarity of the Compare output should be active high or active low, one must understand the difference between the two modes (see Figure 5.8). If the pin is set to the active low state, during the inactive state, the output will be high and low when active. Setting the output to active high will produce opposite results.

As a result, when the output pin is inactive at the beginning of the period, if set to the active low state, the output will be high, and if it is set to the active high state, the output will be low. When a compare event occurs, the state will switch to active resulting in a low output for the active low state and a high output for the active high state. Depending on the state chosen, the pulse of the PWM signal in the active low state will be at the beginning of the period, and in the active high state, it will be at the end of the period. The point being, if a program is designed with the output pin in the active low state and has a duty cycle of 20%, in the active high state, the same program will have a duty cycle of 80%.

The whole program code is in Appendix A – Code #1. In our example, the period and compare values are:

```
EvbRegs.T3PR = 2500;
EvbRegs.CMPR5 = 1250;
```

According to this, we obtain the following PWM signal:

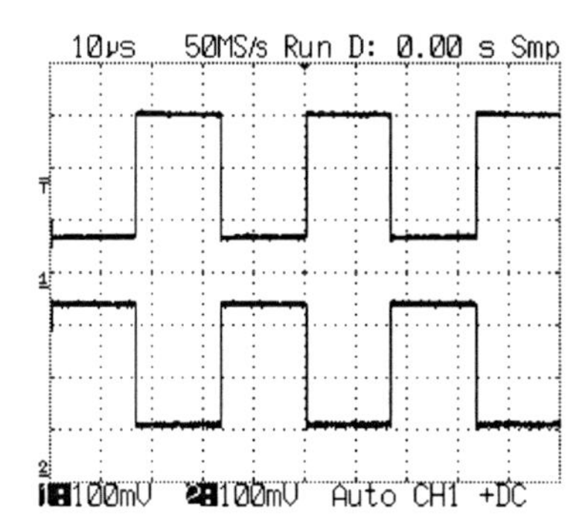

Figure 5.14 PWM Signal with Frequency=30kHz and Duty Cycle=50% at PWM9 and its Complement at PWM10

Furthermore, I connected DIP-switches, which are used to turn on/off the PWM signals, to two of the GPIO pins: T1PWM/T1CMP (Switch #1) and CAP2/QEP2 (Switch #2). In order to achieve this, the following instructions are added in an infinite while-loop to the program:

```
EALLOW;
GpioMuxRegs.GPBMUX.bit.PWM10_GPIOB3 = (GpioDataRegs.GPADAT.bit.GPIOA6+1);
GpioMuxRegs.GPBMUX.bit.PWM9_GPIOB2  =  (GpioDataRegs.GPADAT.bit.GPIOA6+1);
GpioMuxRegs.GPBMUX.bit.PWM8_GPIOB1 = (GpioDataRegs.GPADAT.bit.GPIOA11+1);
EDIS;
```

Switch 1# is linked with "GpioDataRegs.GPADAT.bit.GPIOA6". Therefore, you can turn off the PWM signals PWM9 and its complement PWM10 by closing switch #1. Switch #2 is linked with "GpioDataRegs.GPADAT.bit.GPIOA11" and controls PWM8. Three LED's are connected through buffers to the PWM pins in order to show the PWM states. A photograph of this set-up is shown below:

Figure 5.15 The Experimental Board

5.5 Creating a PWM Signal with Variable Duty Cycle and Frequency

In this chapter, it is explained, how to vary the frequency and duty-cycles of the created PWM signals. You can make it either through keyboard input or analog input. Let us firstly consider the input with the keyboard. In the Code Composer Studio, you have the possibility to change and view the register values. In order to do this, you must open a watch window (in the menu-bar click on GEL and watch registers).

The whole program code is in Appendix A – Code #2. The following additional instructions within the while loop in the program make it possible, to change the frequency, duty cycle and dead-band through variables:

```
EvbRegs.CMPR5 = EvbRegs.T3PR*(DutyCycle/100);
EvbRegs.T3PR = 625*(120/Frequency);
EvbRegs.DBTCONB.bit.DBTPS = DeadBandLarge;
EvbRegs.DBTCONB.bit.DBT = DeadBandFine;
```

After opening the watch window, the variable values can be changed through simple keyboard input. By doing this, you don't have to calculate the needed register values since this is already programmed {625*(120/Frequency)}.

Name	Value	Type	Radix
Frequency	30.0	float	float
DutyCycle	50.0	float	float
DeadBandLarge	5	int	dec
DeadBandFine	10	int	dec

Watch Locals | Watch 1

Figure 5.16 Register Watch Window in CCS

In order to enable the deadband, following instruction must be added to the program:

```
EvbRegs.DBTCONB.bit.EDBT2=1;
```

Here, DeadBandLarge goes from 0 to 5 and DeadBandFine goes from 0 to 15. The default values of the clocking registers are:

HISPCP=0x0001 PLLCR=0x000A

Some examples of possible PWM signals are given in the figures below:

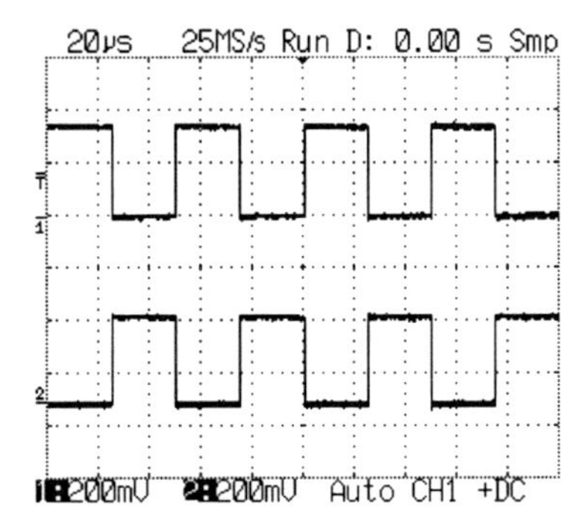

Figure 5.17 PWM with F=20kHz, Duty Cycle=50%, without Dead-Band

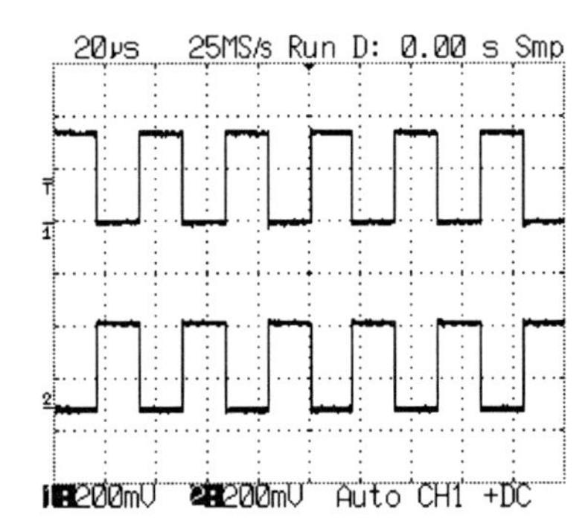

Figure 5.18 PWM with F=30kHz, Duty Cycle=50%, without Dead-Band

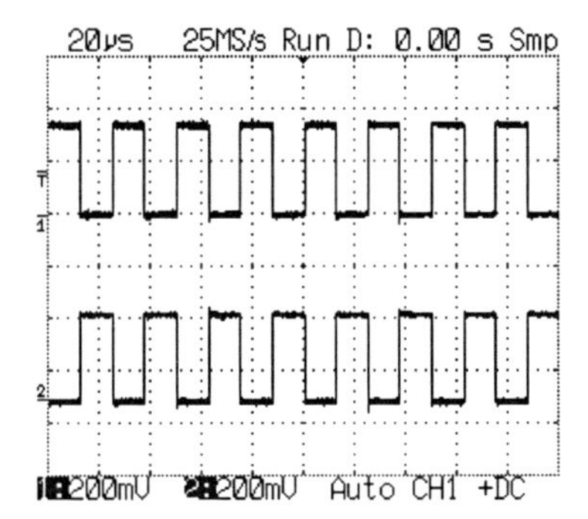

Figure 5.19 PWM with F=40kHz, Duty Cycle=50%, without Dead-Band

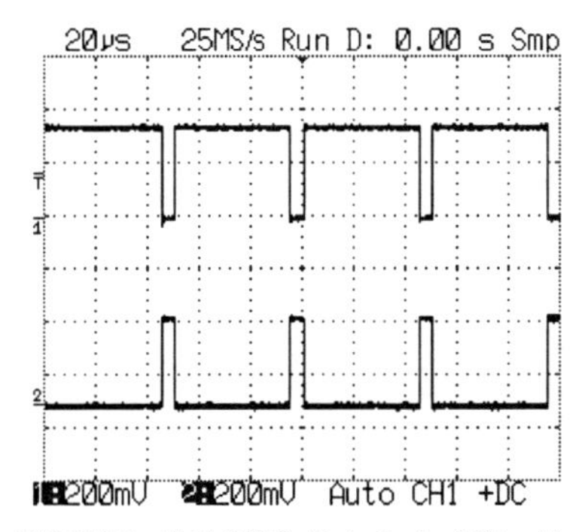

Figure 5.20 PWM with F=20kHz, Duty Cycle=90%, without Dead-Band

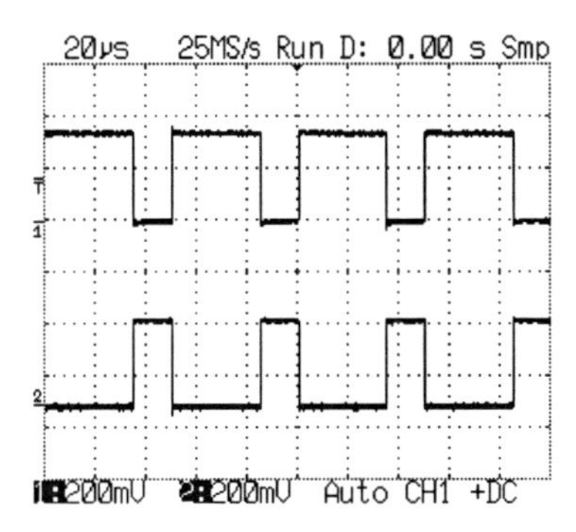

Figure 5.21 PWM with F=20kHz, Duty Cycle=70%, without Dead-Band

Some possibilities for the dead-time settings are given in the table below:

DeadBandLarge	DeadBandFine	Dead-Band Time
5	15	6,4µs
5	14	6,0µs
5	13	5,6µs
5	12	5,2µs
5	11	4,8µs
5	10	4,4µs
5	9	4,0µs
5	8	3,6µs
5	7	3,2µs
5	6	2,8µs
5	5	2,0µs
5	4	1,6µs
5	3	1,2µs
5	2	0,8µs
5	1	0,4µs

Table 5.1 Deadband Register Settings for Dead-Band Generation

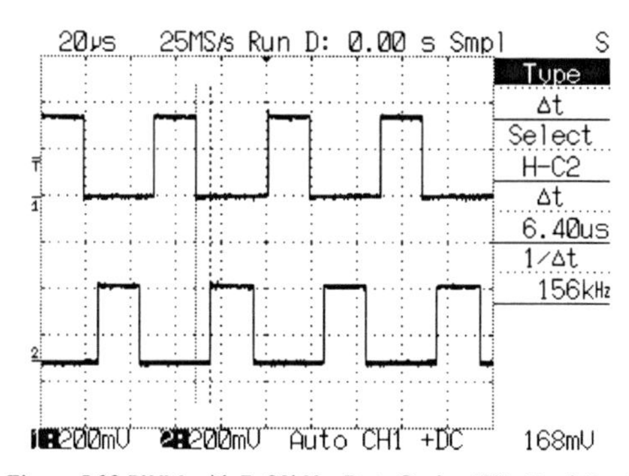

Figure 5.22 PWM with F=20kHz, Duty Cycle=50%, Dead-Band=6.4μs

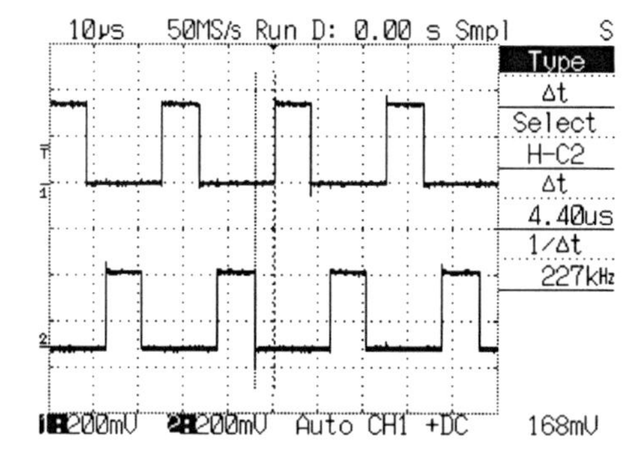

Figure 5.23 PWM with F=20kHz, Duty Cycle=50%, Dead-Band=4.4μs

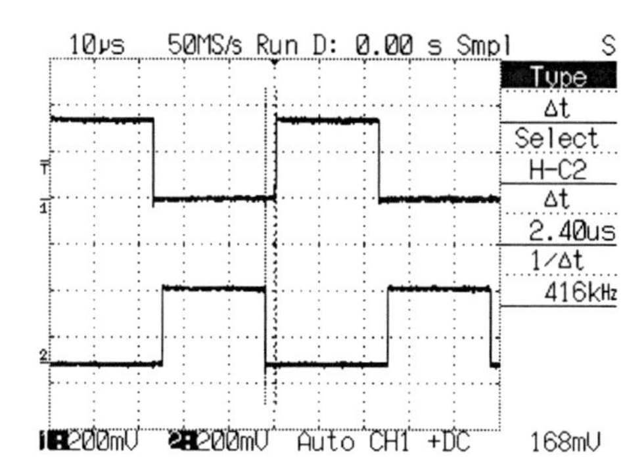

Figure 5.24 PWM with F=20kHz, Duty Cycle=50%, Dead-Band=2.4μs

Up to 448 μs dead-time is possible. For this, the two clocking registers HISPCP and PLLCR should be set as follows:

HISPCP=0x0007 (maximum)
PLLCR=0x0001 (minimum)

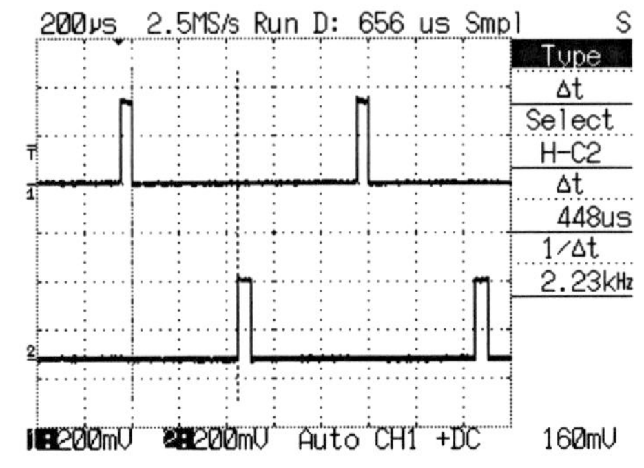

Figure 5.25 PWM with F=1kHz, 50% Duty Cycle and Dead-Band=448μs

In order to run the program without errors, the following source and GEL-files must be added to the program:

Source Files: DSP28_Adc.c
DSP28_DefaultIsr.c
DSP28_GlobalVariableDefs.c
DSP28_PieCtrl.c
DSP28_PieVect.c
DSP28_SysCtrl.c
DSP28_usDelay.asm

GEL Files: f2812.gel
f2812_peripheral.gel

6. Applications

6.1 Creating a Sine Modulated PWM Signal

This application generates an asymmetrical pulse width modulated (PWM) signal with a varying duty cycle. The duty cycle is modulated with a sine function that can be varied in frequency. The implementation of the sine wave modulation is through a look-up table. This application is implemented using C/C++ code. The whole program code is in Appendix A – Code #3.

In Figure 6.1 is a similar application depicted in which the duty cycle is modulated with a triangular function.

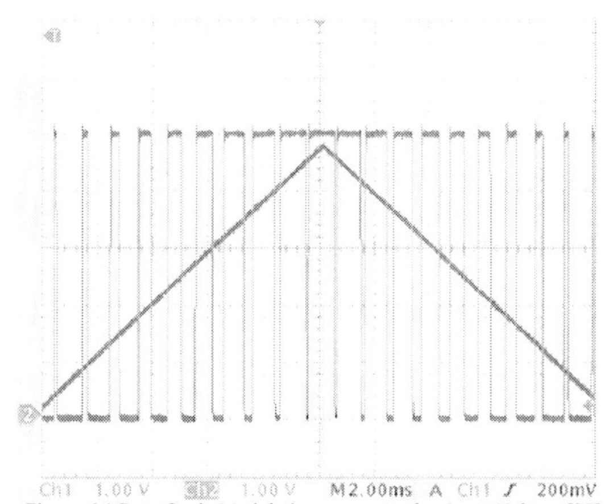

Figure 6.1 Duty Cycle Modulation versus Analog Input Voltage [27]

The implementation of the sine wave modulated PWM signal is simply a modification of the previous application except that the compare registers are modified periodically instead of being held constant.

The generation of the sine wave is performed using a separate file, in which the sine values are stored in.

To determine the frequency of the sine wave, determine how often the value in the compare register will be modified.

Some plots of the sine-modulated PWM:

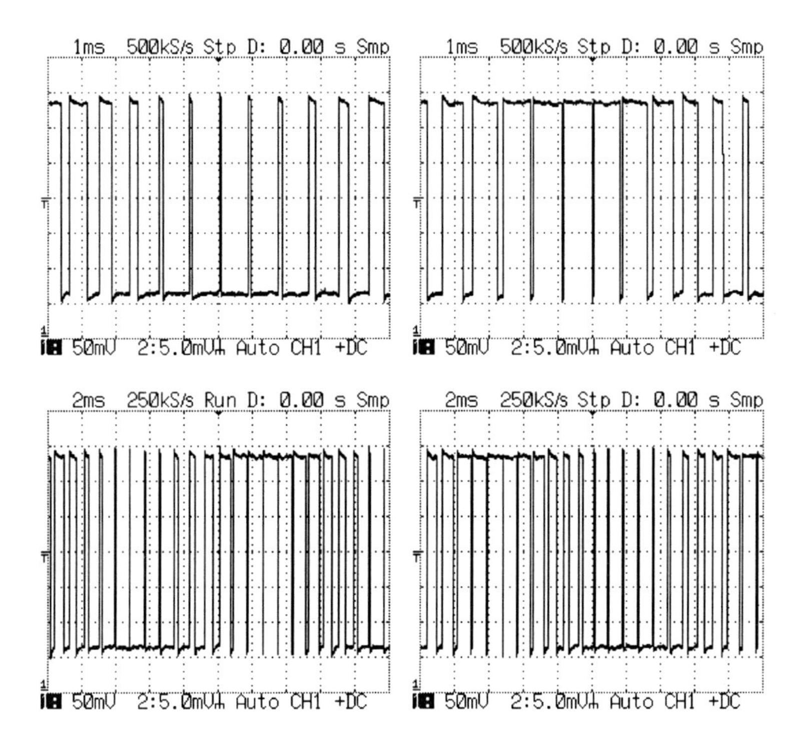

Figure 6.2 Sine-Modulated PWM Signal

To read the sine values from a file, a probe-point must be added to the dataIO function. This is done by clicking on the symbol "toggle probe point":

Then, the file must be added by choosing the menu item File I/O. In this program, the following settings are made:

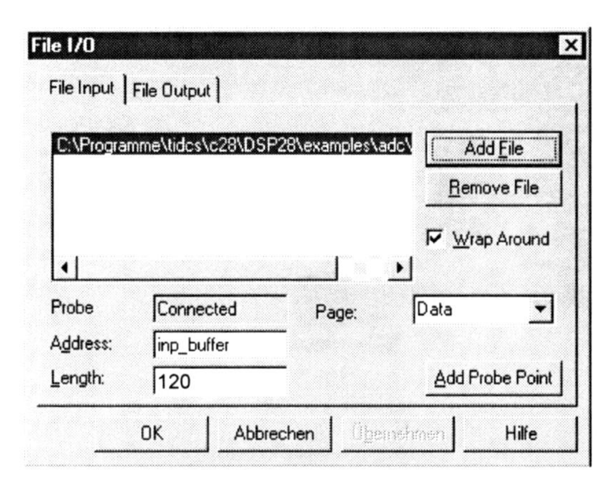

Figure 6.3 Adding a File to the Probe

As address is inp_buffer given, with length 120. That means, the program begins with the first value and read the values one by one until the 120th value. Wrap around means that the program continues from the beginning when the end of file is reached.

To connect the probe point to the file, first mark the probe point, choose the file in the field "Connect To" and click "Replace" as shown below:

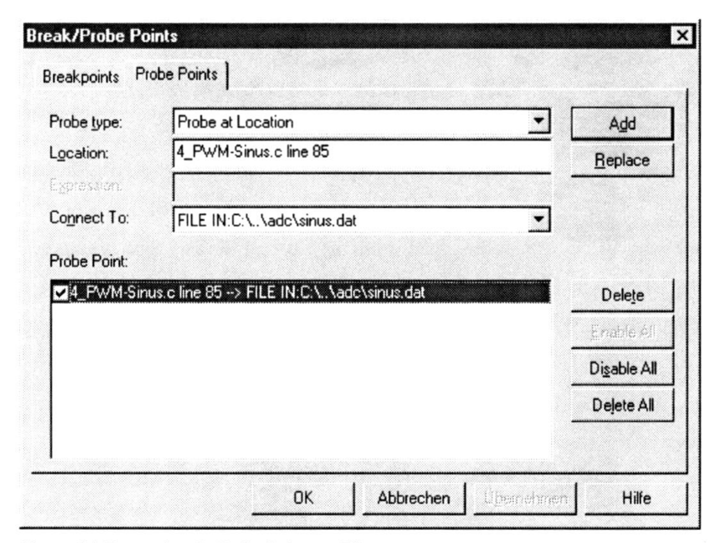

Figure 6.4 Connecting the Probe Point to a File

It is also possible to plot the sine values, which are stored in file. In order to do this, select view/graph in the menu bar and click on time/frequency. In the properties box, type as address again inp_buffer, as length 120 and as the value format 16-bit signed integer. The following graph shows the resulting plot:

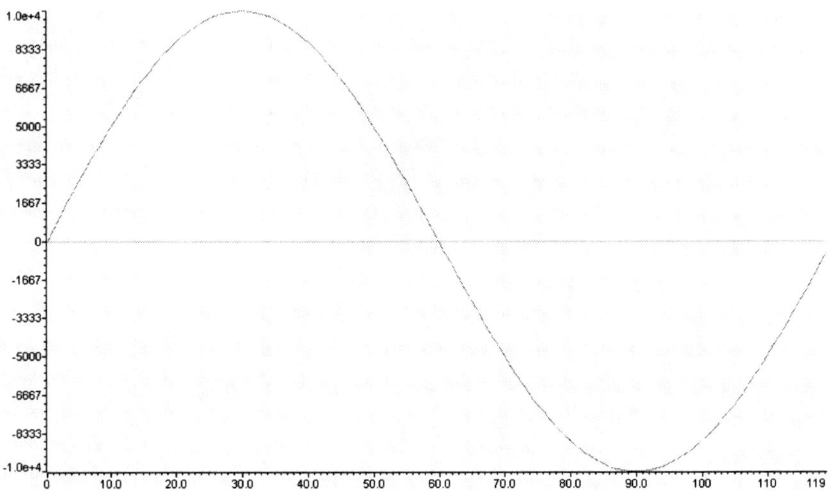

Figure 6.5 Plot of the Stored Sine Values in File

6.1.1 Sine Modulated PWM Generation to Control Inverters

In this program, a sine modulated PWM signal with variable switching frequency, and modulation degree is generated. Also, the sine function can be varied in frequency. The sine values are read through a probe point from an input file. The sine frequency and the modulation degree can be varied through potentiometers. There are four possible values for the switching frequency according to the states of the switches DIP#1 and DIP#2. In this application, the dead-band is 0.5μs and remains constant. This application is implemented using the C/C++ code. The whole program code is in Appendix A – Code #4. The functions of this program can be summarized as follows:

Potentiometer #1 (sine frequency f_1): $20 \text{ Hz} < f_1 < 200 \text{ Hz}$

Potentiometer #2 (modulation degree m): $0 < m < 1$

DIP Switches (switching frequency f_2): DIP #1 and DIP #2

DIP#1	DIP#2	f_2
0	0	2.5 kHz
0	1	5 kHz
1	0	10 kHz
1	1	20 kHz

Table 6.1 States of DIP Switches in Prg a

The photograph of the set-up is shown below:

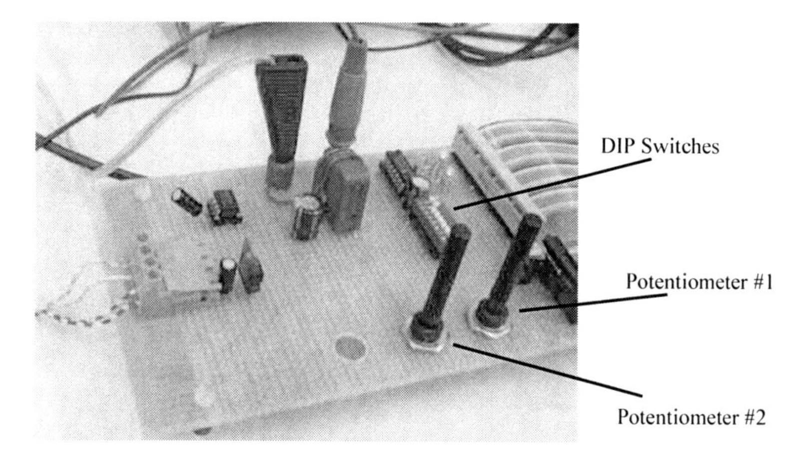

Figure 6.6 The Potentiometers and DIP Switches on the Board

In order to run the program without errors, the following source and GEL-files must be added to the program:

Source Files: DSP28_Adc.c
DSP28_DefaultIsr.c
DSP28_GlobalVariableDefs.c
DSP28_PieCtrl.c
DSP28_PieVect.c

DSP28_SysCtrl.c

DSP28_usDelay.asm

GEL Files: f2812.gel

f2812_peripheral.gel

If not so, then the BUFSIZE value in volume.h should be changed to 120 (number of sine values contained in the input file). Furthermore, the first line in the input file (sinus.dat) must contain '1651 1 0 1 0' in order to be able to read the file.

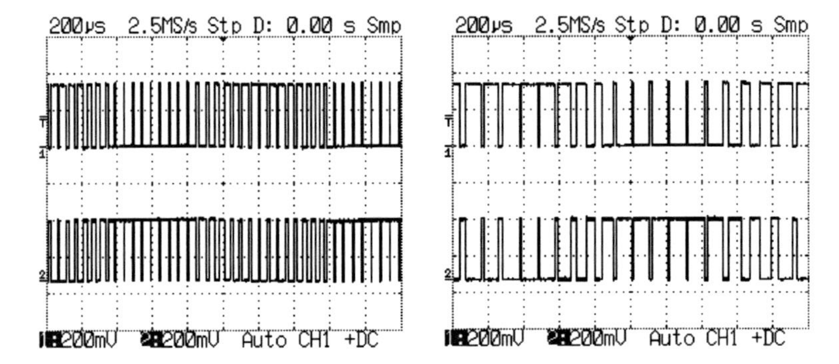

Figure 6.7 Sine Modulated PWM to Control Inverters

6.2 Control of a Half-Bridge of a Switched Mode Power Supply

This EVM application converts two input values with the analog-to-digital converter (ADC) module and outputs a pulse width modulated (PWM) signal corresponding to the digital conversion of the analog inputs. In other words, the pulse width and the frequency of the PWM signal will be proportional to the input values and represent an equivalent analog output signal. The PWM signal will be created asymmetrically. Dead-time is set by using the DIP switches. This application is implemented using the C/C++ code. The whole program code is in Appendix A – Code #5. The functions of this program can be summarized as follows:

Potentiometer #1 (switching frequency f_l): 20 Hz $< f_l <$ 100 kHz
Potentiometer #2 (duty cycle a): $0 < a < 1$
DIP Switches (dead-time): DIP #1 and DIP #2

DIP#1	DIP#2	Deadtime
0	0	0.5 µs
0	1	1 µs
1	0	2 µs
1	1	3 µs

Table 6.2 States of DIP Switches in Prg b

Potentiometer #1 is connected to ADCINA7 and potentiometer #2 is connected to ADCINA2. In order to run the program without errors, the following source and GEL-files must be added to the program:

Source Files: DSP28_Adc.c
DSP28_DefaultIsr.c
DSP28_GlobalVariableDefs.c
DSP28_PieCtrl.c
DSP28_PieVect.c
DSP28_SysCtrl.c
DSP28_usDelay.asm

GEL Files: f2812.gel
f2812_peripheral.gel

Following plots show some PWM signals for different cases of this application:

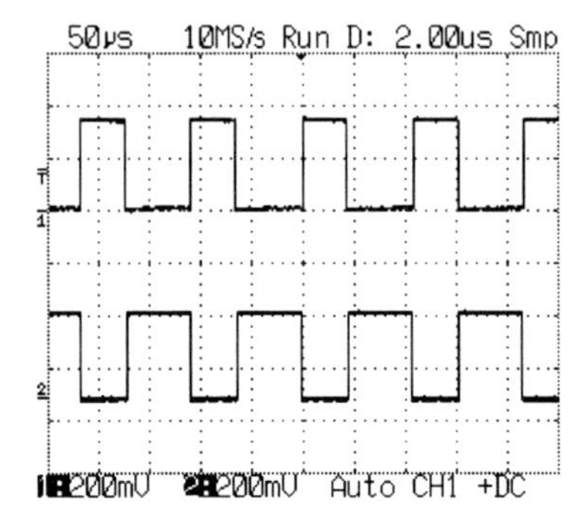

Figure 6.8 PWM with F=9.5kHz, DIP#1DIP#2=01

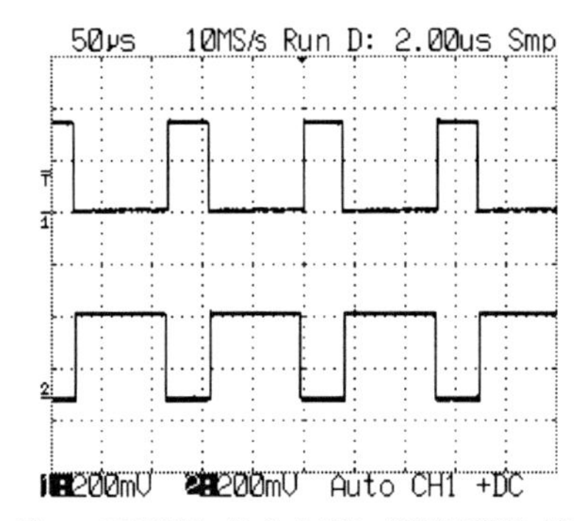

Figure 6.9 PWM with F=7.4kHz, DIP#1DIP#2=10

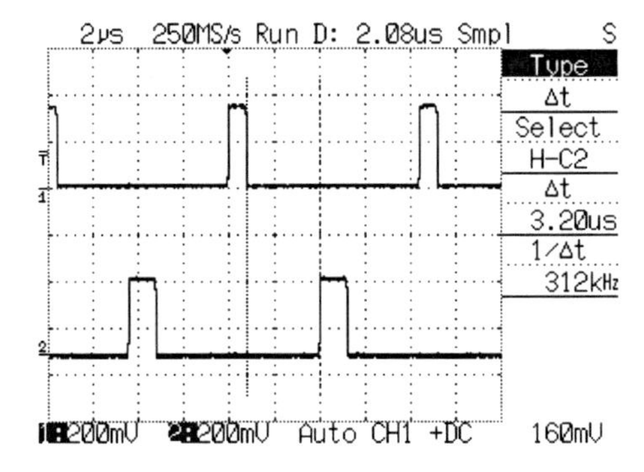

Figure 6.10 PWM with F=120kHz, DIP#1DIP#2=11

6.3 Control of a Series Resonant DC-DC Converter

In many power applications DC voltages have to be converted to an other galvanic isolated voltage level. In the simplest case, DC voltages can be converted with a capacitive voltage divider and a half-bridge to a square wave. This square wave is then transferred through the transformer to the secondary side, whereas the form of the square wave remains constant and its amplitude is set according to the transfer ratio of the windings. Then, at the secondary side only a full bridge rectifier with smoothing capacitor is needed. Hereby, the output voltage is created according to the square wave:

$$U_{out} = \frac{N2}{N1} \cdot \frac{1}{2} \cdot U_{in} \qquad (6\text{-}1)$$

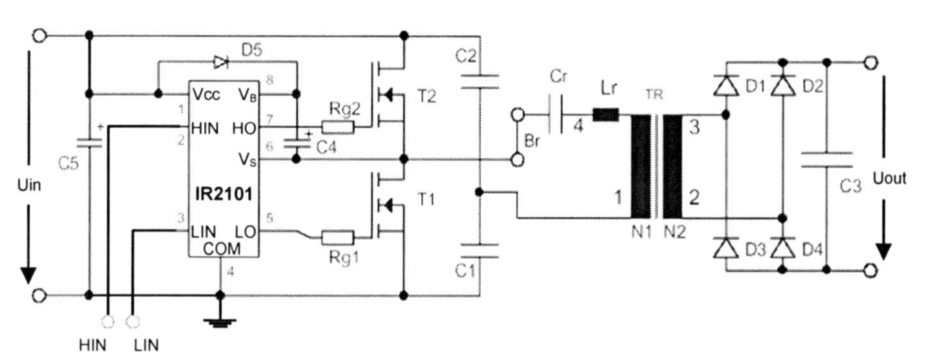

Figure 6.11 The Converter Circuit

This converter operates satisfactorily at low frequencies (e.g. 50Hz). But in this case, a very big transformer would be needed. Since the size of a transformer depends on the frequency, the operating frequency can be increased (e.g. to f ≥ 20 kHz). By doing this, not only the transformer becomes smaller, but also there aren't arising any noises.

But with an increased frequency, a new problem arises. Because of the leakage inductance, a voltage drop occurs in the transformer. Its order of magnitude is at a low percentage in the case of a 50Hz-transformer. Since the corresponding factor for that is $Iac \cdot 2 \cdot \pi \cdot f \cdot L\sigma$, it is obvious that the voltage drop becomes greater with an increasing switching frequency. Consequently, the output voltage and thus the transferable power become smaller. It will be more critical, if because of the good isolation between the primary and secondary side, a high leakage inductance exists.

The solution to this problem is the series resonant converter, which is created by connecting a resonant capacitor C_r in series. By adjusting the switching frequency to the resonant frequency, the optimal operation of the converter is achieved. The DC offset at the leakage inductance is fully compensated through the offset at the capacitor. A sinus-shaped current comes into being, which is in phase with voltage. Hereby, the current is at the moment of switching equal to zero, so that no losses occur (ZCS). The transfer ratio corresponds again to the ideal converter, which normally operates optimally at low frequencies.

6.3.1 The Series Resonant DC-DC Converter

Due to higher efficiency, lower electromagnetic interference and utilization of parasitic inductance and capacitance of power stage components, Resonant Converter topologies are popular in research and industrial and commercial applications.

6.3.1.1 SRC Operation Principle

Figure 6.11 shows the schematics of the power circuit of the Series Resonant DC-DC Converter. The circuit operation consists in closing the pairs of switches Q_1, Q_2 and Q_3, Q_4 alternatively at a frequency above the resonant frequency of the resonant circuit composed by L_r and C_r.

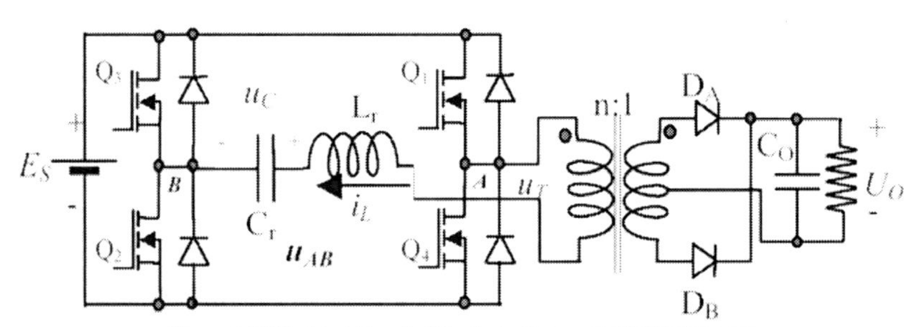

Figure 6.12 Electrical Circuit of the Series Resonant DC-DC Converter

Operation may be obtained at a switching frequency below or above the resonant frequency, however it is desirable to operate above the resonant frequency in order to reduce the switching losses [28]. Operation above the resonant frequency produces an inductive behavior of the circuit, generating a lagging alternating current i_L, in the resonant circuit. Providing that the output capacitor C_O is sufficient to consider that the output voltage does not change significantly, during a few periods of operation, the action of the rectifier diodes D_A and D_B impose at the primary terminals of the power transformer, an alternating square wave voltage u_T, synchronous with the resonant current, with an amplitude of nU_O. The SRC is, therefore, equivalent to an L_rC_r circuit supplied by two alternating square voltages u_{AB} (generated by the action of the switches Q_1, Q_2, Q_3, and Q_4) with an amplitude equal to the input voltage, and voltage u_T (generated by the action of the output diodes D_A and D_B), as presented in Figure 6.12 [29].

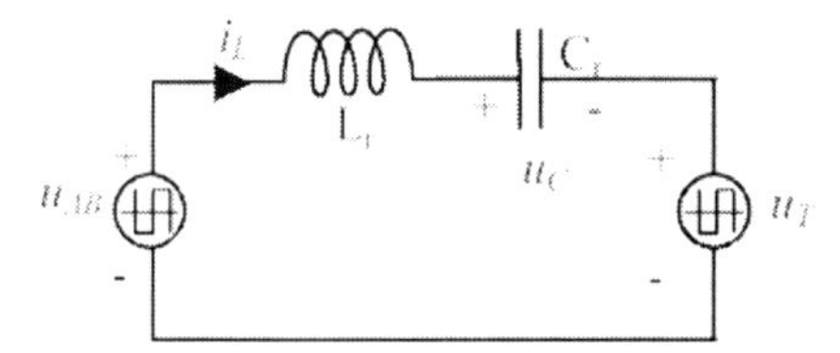

Figure 6.13 Series Resonant DC-DC Converter Equivalent Circuit

The converter has four modes of operation defined by the four combinations of the values of the two power supplies. Table 6.3 identifies the operation modes and the correspondent conducting switches and generated voltages.

Voltages \ Mode	I	II	III	IV
u_{AB}	$+E_S$	$-E_S$	$-E_S$	$+E_S$
U_T	NU_O	nU_O	$-nU_O$	$-nU_O$
U_{LC}	E_S-nU_O	$-E_S-nU_O$	$-E_S+nU_O$	E_S+nU_O
Switch on	Q_1, Q_2, D_A	D_3, D_4, D_A	Q_3, Q_4, D_B	D_1, D_2, D_B

Table 6.3 SRC Operating Modes

In most cases series resonant power circuits show high quality factors. Hence, the operation point of a converter strongly depends on the ratio of the control frequency to the resonant frequency. This may cause problems for the control if the resonant frequencies change according to the value tolerances of the capacitors or because of movable coils (L_r changes) for contactless power transfer [30].

The problem can be solved by a fast control algorithm presented in this chapter. It estimates the unknown or variable resonant frequency of a series resonant converter by successive measurement of the converter current at certain time instants within a conduction period. As a result the converter frequency can continuously follow the resonant frequency of the system.

The SRC circuit consists of a full bridge transistor inverter, a series capacitor C_r, a HF transformer and a secondary rectifier with smoothing capacitor C_O. The resonant capacitor is inserted on the primary side in order to prevent a DC offset at the transformer. The resonant frequency f_R of the converter is mainly determined by the leakage inductance of the transformer L_r and the capacitance C_r of the series capacitor [30]:

$$f_R = \frac{1}{2 \cdot \pi \cdot \sqrt{L_r \cdot C_r}} \tag{6-2}$$

In order to vary the leakage inductance of the transformer the secondary winding coil can be moved on the core while the primary winding coil is fixed. The transformer can be seen in Figure 6.13.

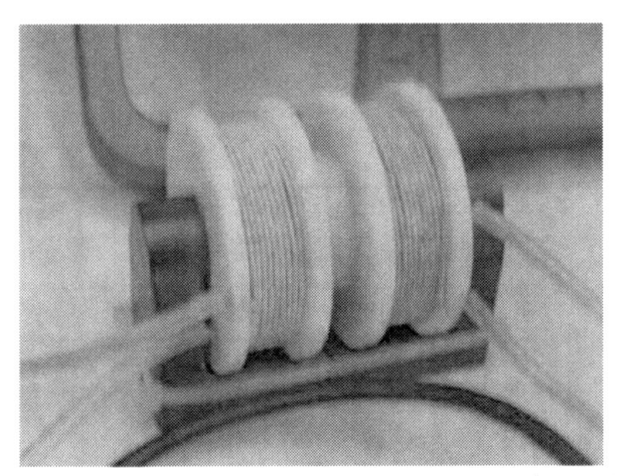

Figure 6.14 Transformer Set-Up with a Fixed Primary Coil and a Movable Secondary Coil

The transistor inverter is controlled by the PWM unit of a DSP, which contains a fast A/D converter. The DSP allows frequency and duty cycle setting as well as sample-and-hold

(S&H) operation at defined, programmable time instants, which are synchronized to the switching operation.

The converter current I_c can be sampled in equidistant time steps T as illustrated in Figure 6.14.

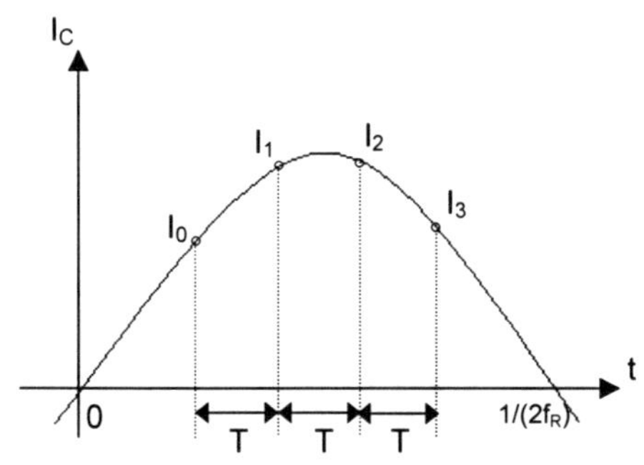

Figure 6.15 Sampling of the Sinusoidal Current

In this control strategy, at least 3 measured values are required (N=3):

$$I_0 = \hat{I} \cdot \sin(2 \cdot \pi \cdot f_R \cdot (t_0)) \tag{6-3}$$

$$I_1 = \hat{I} \cdot \sin(2 \cdot \pi \cdot f_R \cdot (t_0 + T)) \tag{6-4}$$

$$I_2 = \hat{I} \cdot \sin(2 \cdot \pi \cdot f_R \cdot (t_0 + 2 \cdot T)) \tag{6-5}$$

Here, the sampling distance T and the measured values I_0, I_1 and I_2 are known. Unknown values are $\hat{I}$, t_0 and f_R.

Now, if we add I_0 and I_2, we obtain:

$$I_0 + I_2 = \hat{I} \cdot [\sin(\omega \cdot t_0) + \sin(\omega \cdot t_0 + 2 \cdot \omega \cdot T)] \tag{6-6}$$

where

$$\omega = 2 \cdot \pi \cdot f_R \tag{6-7}$$

according to the formula

$$\sin(a) + \sin(b) = 2 \cdot \sin(\frac{a+b}{2}) \cdot \cos(\frac{a-b}{2}) \tag{6-8}$$

$I_0 + I_2$ becomes:

$$I_0 + I_2 = 2 \cdot \hat{I} \cdot [\sin(\omega \cdot t_0 + \omega \cdot T) \cdot \cos(-\omega \cdot T)] \quad (6\text{-}9)$$

since

$$\cos(-a) = \cos(a) \quad (6\text{-}10)$$

$$\Rightarrow I_0 + I_2 = 2 \cdot \underbrace{\hat{I} \cdot [\sin(\omega \cdot t_0 + \omega \cdot T)}_{= I_1} \cdot \cos(\omega \cdot T)]$$

$$\Rightarrow I_0 + I_2 = 2 \cdot I_1 \cdot \cos(\omega \cdot T) \quad (6\text{-}11)$$

$$\Rightarrow \cos(\omega \cdot T) = \frac{I_0 + I_2}{2 \cdot I_1} \quad (6\text{-}12)$$

$$\Rightarrow \omega \cdot T = ar\cos\left(\frac{I_0 + I_2}{2 \cdot I_1}\right) \quad (6\text{-}13)$$

since $\omega = 2 \cdot \pi \cdot f_R$

$$\Rightarrow 2 \cdot \pi \cdot f_R \cdot T = ar\cos\left(\frac{I_0 + I_2}{2 \cdot I_1}\right) \quad (6\text{-}14)$$

From this, it follows that $f_R = \frac{1}{2 \cdot \pi \cdot T} \cdot ar\cos\left(\frac{I_0 + I_2}{2 \cdot I_1}\right)$ (6-15)

Now, the frequency f_R can be estimated by using the measured current values I_0, I_1 and I_2 according to this formula:

$$f_R = \frac{1}{2 \cdot \pi \cdot T} ar\cos(\rho) \quad (6\text{-}16)$$

where

$$\rho = \frac{I_0 + I_2}{2 \cdot I_1} \quad (6\text{-}17)$$

If N=4 sample values are taken it results in [30]

$$\rho = \frac{I_1 \cdot (I_0 + I_2) + I_2 \cdot (I_1 + I_3)}{2 \cdot (I_1^2 + I_2^2)} \quad (6\text{-}18)$$

The resolution of $2.\pi.f_R.T$ becomes rather poor for larger values of ρ near one. By increasing N usually a higher accuracy can be obtained for ρ. However, the sampling distance T decreases accordingly. This leads to larger values for ρ which increases the instability of equation due to the infinite slope of the arcos function at $\rho=1$. Thus, a small number of points are favorable for measurement. It also limits the processing effort [30].

The whole program code is in Appendix A – Code #6. In order to run the program without errors, the following source and GEL-files must be added to the program:

Source Files: DSP28_Adc.c
DSP28_DefaultIsr.c
DSP28_GlobalVariableDefs.c
DSP28_PieCtrl.c
DSP28_PieVect.c
DSP28_SysCtrl.c
DSP28_usDelay.asm

GEL Files: f2812.gel
f2812_peripheral.gel

The following three figures on the next pages show the converter current and PWM signal for three cases:

Case #1: The frequency of the converter current is higher than the switching frequency:

25 kHz = Converter Frequency > 20 kHz = Switching Frequency

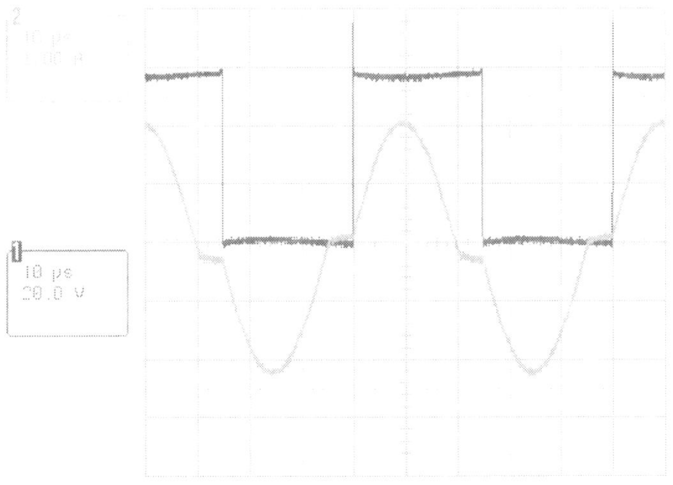

Figure 6.16 Converter Frequency is Higher than the Switching Frequency

Case #2: The frequency of the converter current is equal to the switching frequency:

Converter Frequency = Switching Frequency = 20 kHz

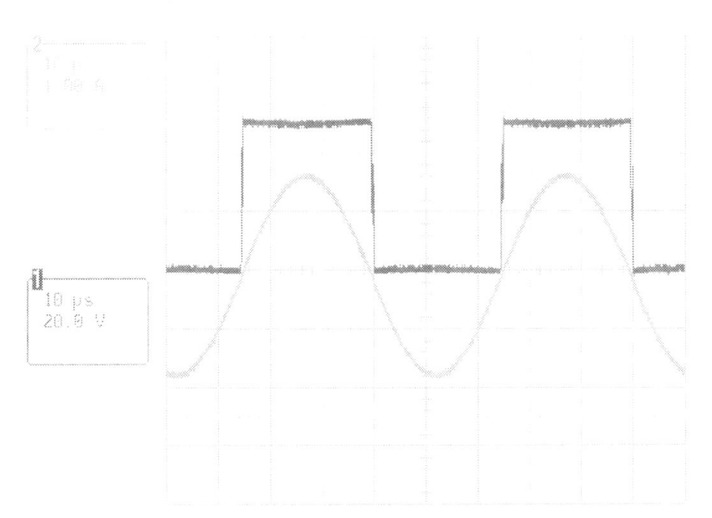

Figure 6.17 Converter Frequency is Equal to the Switching Frequency

Case #3: The frequency of the converter current is lower than the switching frequency:

18.5 kHz = Converter Frequency < 23.8 kHz = Switching Frequency

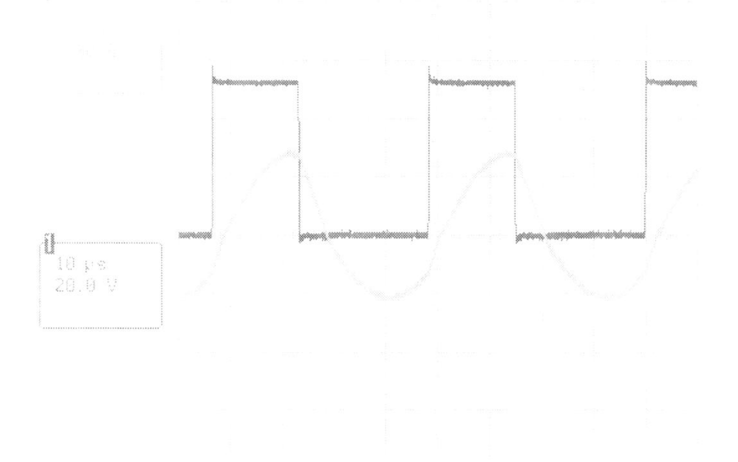

Figure 6.18 Converter Frequency is Lower than the Switching Frequency

As it can be seen from the figures above, the current must be measured always in a period, in which it is sinus-shaped. The measurements occur at a suitable constant sampling distance T. Moreover, only positive values are used for the estimation and finally, the frequency is calculated as a mean-value of every 20 period-values in order to obtain a stable result for the frequency.

6.3.2 The Snubber Effect

The series resonant DC-DC converter sets the output voltage U_{out} to a value that is in a fixed relation to the input voltage U_{in} so far as it is operated with the resonant frequency $f_s = f_{res}$ [56]. It is the optimal solution if a fixed transfer ratio is sufficient.

In some applications however, it is desired to have an adjustable output voltage while the input voltage remains constant. But an adjustment of the output voltage is only possible, if the frequency is varied [56]. For example, at a frequency, which is greater than the resonant frequency, we can expect that the output voltage is getting smaller. But hereby, a phase shift occurs between current and voltage, that the voltage is leading and at the moment of switching on, the current is unequal zero. Consequently, switching losses occur. There are two possible cases for the switching frequency:

a) Operation at $f_{min} < f_s < f_{res}$: Current will be switched on with transistor (high losses)
b) Operation at $f_{res} < f_s < f_{max}$: Current will be switched off (ZVS is possible)

For the operation with $f_s > f_{res}$, the switching losses can be reduced by making the rising time of the voltage greater. This can be done by adding a capacitor in parallel to the transistor. Now, after switching off the one transistor, if the other complementary transistor won't be switched on directly, so in the meantime the positive transformer current can slowly load the capacitor. When the input voltage is reached, the current can flow through the intern diode of the transistor and the voltage will be jammed. Subsequently, the complementary transistor can be switched on without any voltage. Thus, there are arising no switching losses (ZVS).

The use of either lossless snubber circuits or resonant zero-voltage-switching and zero-current-switching (ZVS and ZCS) techniques increases the efficiency of power converters and inverters, and reduces electromagnetic interference (EMI) and noise caused by stray inductance, parasitic capacitance, and other imperfections in practical circuits and devices [48,49]. In the case of a snubber, the switch current increase is delayed at turn-on and the switch voltage rising is delayed at turn-off, avoiding an overlap of supply voltage and load current in the switching period. Passive components are normally used, although lossless snubbers incorporate complications associated with energy recovery [50,51,52]. In contrast, with ZVS the switch voltage is forced to zero before the switch current rises at turn-on. In the case of ZCS the switch current is forced to zero before the switch voltage rises at turn-off [53].

To show the snubber effect, a capacitor Cs = 22nF is connected parallel to transistor T1. The modified circuit is shown on the next page.

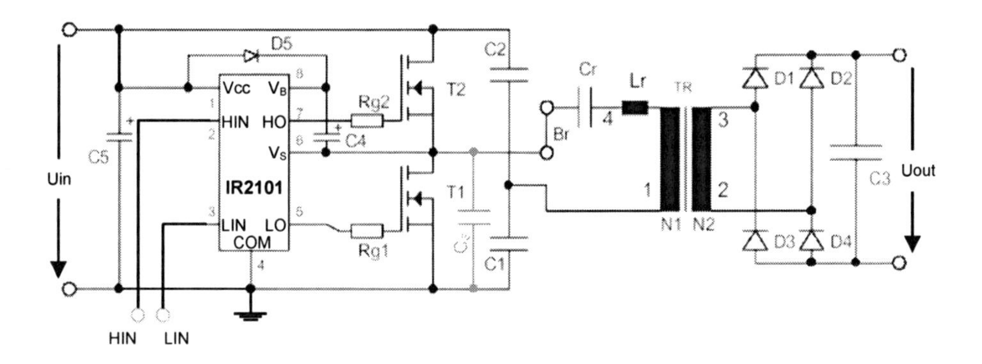

Figure 6.19 SRC Circuit with Snubber Capacitor C_S Added Parallel to T1

For $f > f_{res}$ is $I_c(t) = I_{in}$ at the moment of switching on. This current can be used to commutate the transistor voltage linear from 0 to U_{in}. As it can be also seen from Figure 6.19, the dead time t_d must be greater than the commutating time of U_{T1}, t_c. In our case, t_c is about 1.5μs (see Fig. 6.20). Therefore, dead-time must be greater than 1.5μs.

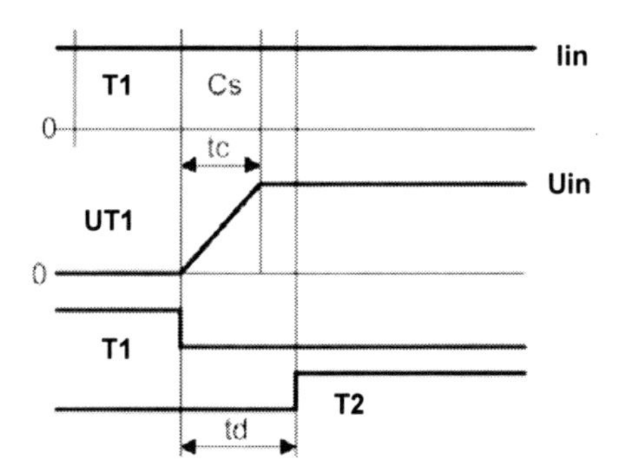

Figure 6.20 Commutating the Transistor Voltage linearly from 0 to U_{in} [56]

This commutating procedure is also called soft-switching and it doesn't depend on the input voltage but requires that the commutating time tc is smaller than the dead-time td. The series resonant DC-DC converter that is optimized by soft-switching combines the advantages of various resonant principles. Hereby, an optimal EMI behavior and a good efficiency is achieved [54].

The following figures show the effect of the snubber capacitor.

1) Operation without snubber capacitor. The voltage rise time is 100ns:

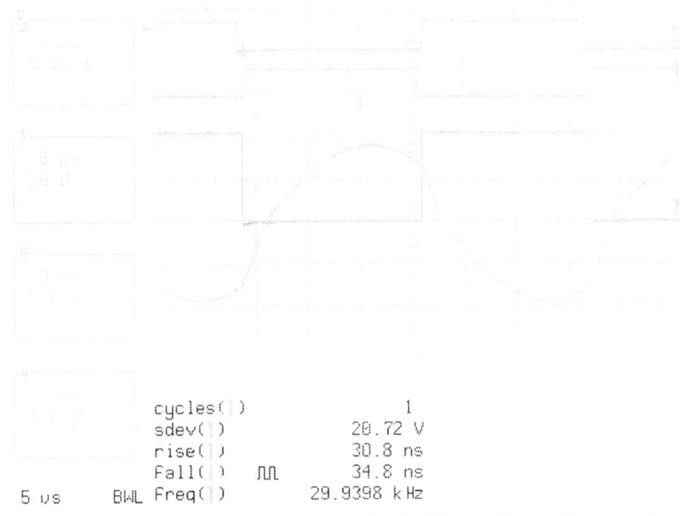

Figure 6.21 UT1(1) and Converter Current(2) when Dead-Time=0.5µs, without Capacitor

2) Operation with snubber capacitor and dead-time is greater than commutating time. The voltage rise time is 1.5µs:

Figure 6.22 UT1(1) and Converter Current(2) when Dead-Time=3µs

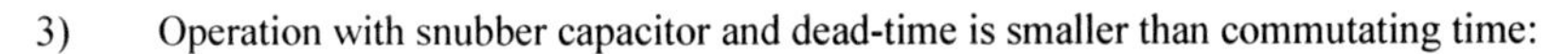

3) Operation with snubber capacitor and dead-time is smaller than commutating time:

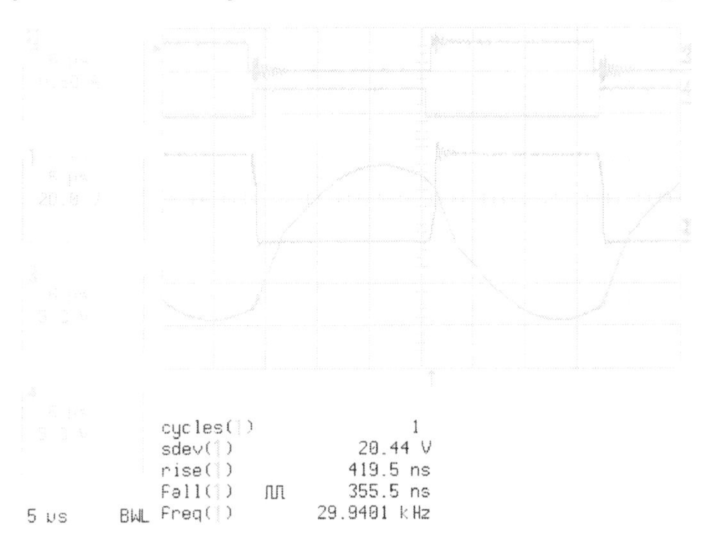

Figure 6.23 UT1(1) and Converter Current(2) when Dead-Time=0.5µs

In view of electromagnetic interference (EMI), slower rising flanks have also less interference.

7. Conclusion and Recommended Continuation

7.1 Conclusion

After a brief introduction on power electronics and control systems, short review of DSPs and digital signal processing, the TMS320F2812 and its corresponding eZdsp board were presented in the first chapters of this thesis. Furthermore, the software development of the DSP and the steps required to create a project, to build and to run a program, were described. A definition of pulse width modulation, which is the focus of this thesis, PWM generation and PWM output configuration principles were given in the next sections.

Program examples, whose codes are in the appendices part, were given in the last chapters of this thesis. Moreover, how to generate PWM signals varied with analog or keyboard inputs or according to a sine wave, whose values are stored in a file, were described. At last, some control applications, e.g. control of a series resonant DC-DC converter, with the use of PWM signals were presented.

7.2 Future Work

After the work and studies that were done following steps should be taken in further study of PWM control and the set-up of the SRC control:

- Other studies should be performed in order to improve the PWM generation using an analog input (with less interference and without the need for truncation)

- In control of the series resonant DC-DC converter, a method should be developed to synchronize the ADC conversion frequency with the varying switching frequency in order to obtain more exact frequency values.

Bibliography

[1] Claus-Ulrich Karipidis. A Versatile DSP/FPGA Structure optimized for Rapid Prototyping and Digital Real-Time Simulation of Power Electronic and Electrical Drive Systems. Doctora Thesis, 2001.

[2] John M. D. Murphy, Fred G. Turnbull. Power Electronic Control of AC Motors. Pergamon Press, 1988.

[3] Werner Leonhard. Control of Electrical Drives. Springer Verlag, 1996.

[4] Ned Mohan, Tore M. Undeland, William P. Robbins. Power Electronics: Converters, Applications and Design. John Wiley and sons, 1989.

[5] R. D. Lorenz, T. A. Lipo, D. W. Nowotny. *Motion Control with Induction Motors.* Proceedings of the IEEE, vol.82, no.8, August 1994.

[6] E. Kiel, F. Profumo, W. Schumacher. *Microprocessor Control of AC Drives.* Tutorial at EPE'95.

[7] B. K. Bose. *Expert System, Fuzzy Logic and Neural Network Applications in Power Electronics and Motion Control.* Proceedings of the IEEE, vol.82, no.8, August 1994.

[8] Cotistantm Ilas, Aurelian Sarca, Radu Giuclea, Liviu Kreindler. *Using TMS320 Family DSPs in Motion Control Systems.* Texas Instruments, 1996.

[9] P. Vas, W. Drury. *Future Trends and Development of Electrical Machines and Drives.* PCIM'95.

[10] Charles L. Phillips, H. Troy Nagle. Digital Control System Analysis and Design. Prentice Hall, 1995.

[11] Toshio Takahashi. *New Digital Hardware Control Method for High Performance AC Servo Motor Drive.* International Rectifier, 2002.

[12] Yongxuan Hu. *Analysis and Design of High-Intensity-Discharge Lamp Ballast for Automotive Headlamp*. November 2001.

[13] H. Matsuo, F. Kurokawa, Z. Luo, Y. Makino, Y. Ishizuka, T. Oshikata. *Partially Resonant Active Filter Using the Digital PWM Control Circuit with the DSP*. IEEE TELESCON, 2000.

[14] Andrew Bateman, Iain Paterson-Stephens. *The DSP Handbook*. Prentice Hall, 2002.

[15] Texas Instruments Inc. http://www.ti.com

[16] R. C. Maher. *A Short Introduction to DSP Microprocessor Architecture*. ECEN4002/5002 DSP Laboratory. Spring 2003.

[17] Artem Doudine, Ilgiz Mukhamedzhanov. *Benchmarking DSP processor, Methods and Implementations*. Linköping, 2003.

[18] Paul Sparacino. *A New TI DSP Product Line*. Texas Instruments, 2002.

[19] *TMS320F28x DSP Peripherals Reference Guide*. Texas Instruments, May 2002.

[20] *TMS320F28x Analog-to-Digital Converter (ADC) Peripheral Reference Guide*. Texas Instruments, June 2002.

[21] *TMS320F28x Event Manager (EV) Peripheral Reference Guide*. Texas Instruments, May 2002.

[22] *eZdsp F2812 Technical Reference*. Spectrum Digital, Inc., July 2002.

[23] David M. Alter. *Using PWM Output as a Digital-to-Analog Converter on a TMS320C240 DSP*. Texas Instruments, November 1998.

[24] Application Report. *SPRA371*. Texas Instruments, 1997.

[25] Mohammed S. Arefeen. *Configuring PWM Outputs of TMS320F240.* Texas Instruments, 1997.

[26] David Figoli. *Creating a Pulse Width Modulated Signal with a Fixed Duty Cycle Using the TMS320F240 EVM.* Texas Instruments, January 1999.

[27] Alfredo Olmos, Andre Vilas Boas, Marcus Serra Espindola. *Generating a PWM Signal Using the MC68HC908QY4 Microcontroller.* Motorola, 2003.

[28] S. Valtchev, J. B. Klaassens. *Efficient Resonant Power Conversion.* IEEE Transactions on Industrial Electronics, vol.37, Dec.1990.

[29] Stanimir Valtchev, Beatriz V. Borges, J. B. Klaassens. *Series Resonant Converter Applied to Contactless Energy Transmission.*

[30] Prof. Dr. Heinz van der Broeck, Christian Hattrup, Martin Ossmann. *Fast Estimation Techniques for Digital Control of Resonant Converters.* IEEE PELS Transactions on Power Electronics, 1. January 2003.

[31] Brian W. Kernighan, Dennis M. Ritchie. *The C Programming Language.* Prentice Hall International, 1989.

[32] Prof. Dr. Heinz van der Broeck. *Simulation of PWM induction motor drives by means of Mathcad.* Lausanne, EPE '99.

[33] Paul M. Embree. *C Algorithms for Real-Time DSP.* Prentice Hall, 1995.

[34] P. Horowitz, W. Hill. *The Art of Electronics.* Cambridge University Press, Second Edition, 1989.

[35] James H. McClellan, Ronald W. Schafer, Mark A. Yoder. *DSP First.* Prentice Hall, 1998.

[36] Richard C. Jaeger. *Microelectronic Circuit Design*. The McGraw-Hill Companies, Inc., 1997.

[37] William Ford, William Topp. *Data Structures with C++*. Prentice Hall, 1996.

[38] FH Zwickau. http://www.fh-zwickau.de/tutorial/dsp

[39] Analog Devices Inc. http://www.analog.com

[40] Motorola Inc. http://www.motorola.com

[41] DSP World Conferences on Digital Processing Solutions and Technologies. http://www.dspworld.com

[42] DSP Group, Inc. http://www.dspg.com

[43] Jigna Patel. *The Hottest Markets for External Power Supplies Now and a Look at Potential Future Markets*. APEC'01.

[44] Mark T. Gaboriault. *U.S. Merchant Markets and Applications for Internal AC/DC Switching Power Supplies and DC/DC Converters*. APEC'00.

[45] P. C. Todd. *UC3854 Controlled Power Factor Correction Circuit Design*. Application Note U-134. Unitrode Corporation/Texas Instruments.

[46] Dr. P. Enjeti, Sangsun Kim. *Digital Control of Switching Power Supply - Power Factor Correction Stage*.

[47] Christian Hattrup. *Ansteuerung von Stromrichtern mit Hilfe des Mikrocontrollers 80C166 und des hierauf basierenden VeCon-Chips*. Diploma Thesis, 1995.

[48] B. Orlik, O. Scheuer. *Optimizing switching losses and EMC of pulse controlled inverters using EMC snubber circuit*. Proc. European Power Electronics and Applications Conf., vol. 4, Sept. 1997.

[49] S. Cazabat, W. Melhem, A. Puzo, J. Gonzalez, F. Forest, R. Critchley, H. Pouliquen. *High power soft switching PWM IGBT converter electrical and EMC characterization.* Proc. European Power Electronics and Applications Conf., vol. 4, Sept. 1997.

[50] W. McMurray. *Efficient snubbers for voltage-source GTO inverters.* IEEE Transactions on Power Electronics, vol. PE-2, July 1987.

[51] J. Holtz, S. Salama, K. H. Werner. *A nondissipative snubber circuit for high-power GTO inverters.* IEEE Trans. Ind. Applicat., vol. 25, July 1989.

[52] D. Tardiff. *A summary of resonant snubber circuits for transistors and GTOs.* Conf. Rec. IEEE-IAS Annu. Meeting, Oct. 1989.

[53] K. Sheng, X. He, B. W. Williams, Z. Qian, S. J. Finney. *A Composite Soft-Switching Inverter Configuration with Unipolar Pulsewidth Modulation Control.* IEEE Transactions on Industrial Electronics, vol. 48, Feb. 2001.

[54] Prof. Dr. Heinz van der Broeck. *Potentialtrennender DC Leistungswandler mit optimierter Schaltentlastung.* Forschungsbericht 1998-2000, FH Köln.

[55] *Schematics of eZdsp F2812 board assembly revision B.* Spectrum Digital, July 2002.

[56] Prof. Dr. Heinz van der Broeck. *Serienresonanzkonverter mit verstellbarer Ausgangsspannung.* Schaltnetzteile II, RWTH Aachen, SS 2003.

Appendix

A. Program Codes

Program Code #1

```
// Creating a PWM Signal with Fixed Duty Cycle and Frequency

// Include required header files
#include "DSP28_Device.h"

void main(void)
{
// Initialize System Control registers, PLL, WatchDog, Clocks to default state:
        InitSysCtrl();

        // Disable and clear all CPU interrupts:
        DINT;
        IER = 0x0000;
        IFR = 0x0000;

        // Initialize Pie Control Registers To Default State:
   InitPieCtrl();

        // Initialize the PIE Vector Table To a Known State:
        InitPieVectTable();

   // Initialize EVB Timer3
        EvbRegs.T3PR = 2500;         // Timer3 period
        EvbRegs.T3CMPR = 0x3C00;         // Timer3 compare
        EvbRegs.T3CNT =  0x0000;         // Timer3 counter

   // TMODE = continuous up/down
        // Timer enable
        // Timer compare enable
        EvbRegs.T3CON.all  = 0x1042;

 // Enable compare
        EvbRegs.CMPR5 = 1250;

   // Compare action control.
   EvbRegs.ACTRB.all = 0x0666;
        EvbRegs.DBTCONB.all = 0x0000; // Disable deadband
   EvbRegs.COMCONB.all = 0xA600;
```

```
 while (1) // Infinite loop.
 {
 EALLOW; // Enable PWM pins.
 // Turn on/off the PWM signals through switches:
 GpioMuxRegs.GPBMUX.bit.PWM10_GPIOB3=(GpioDataRegs.GPADAT.bit.GPIOA6+1);
 GpioMuxRegs.GPBMUX.bit.PWM9_GPIOB2 = (GpioDataRegs.GPADAT.bit.GPIOA6+1);
 EDIS; // Disable PWM pins.
 }

}
```

Program Code #2

```
// Creating a PWM Signal with variable Duty Cycle and Frequency (keyboard input)

// Include required header files
#include "DSP28_Device.h"

float Frequency, DutyCycle;
int DeadBandLarge, DeadBandFine;

void main(void)
{
//  Initialize System Control registers, PLL, WatchDog, Clocks to default state:
        InitSysCtrl();

        // Disable and clear all CPU interrupts:
        DINT;
        IER = 0x0000;
        IFR = 0x0000;

        // Initialize Pie Control Registers To Default State:
   InitPieCtrl();

        // Initialize the PIE Vector Table To a Known State:
        InitPieVectTable();

   // Initialize EVB Timer3
        EvbRegs.T3PR = 2500;        // Timer3 period
        EvbRegs.T3CMPR = 0x3C00;        // Timer3 compare
        EvbRegs.T3CNT =  0x0000;        // Timer3 counter
   // TMODE = continuous up/down
        // Timer enable
        // Timer compare enable
        EvbRegs.T3CON.all  = 0x1042;

 // Enable compare
        EvbRegs.CMPR5 = 1250;

   // Compare action control.
   EvbRegs.ACTRB.all = 0x0666;
        EvbRegs.DBTCONB.bit.EDBT2 = 1;        // Enable deadband
   EvbRegs.COMCONB.all = 0xA600;
```

```
while (1) // Infinite loop.
 {
 EALLOW; // Enable PWM pins.
 // Turn on/off the PWM signals through switches:
 GpioMuxRegs.GPBMUX.bit.PWM10_GPIOB3=(GpioDataRegs.GPADAT.bit.GPIOA6+1);
 GpioMuxRegs.GPBMUX.bit.PWM9_GPIOB2 = (GpioDataRegs.GPADAT.bit.GPIOA6+1);
 EDIS; // Disable PWM pins.

 // Binding of duty cycle to compare register:
 EvbRegs.CMPR5 = EvbRegs.T3PR*(DutyCycle/100);
 // Binding of frequency to period register:
 EvbRegs.T3PR = 625*(120/Frequency);
 // Binding deadband to DBTCONB registers:
 EvbRegs.DBTCONB.bit.DBTPS = DeadBandLarge;
 EvbRegs.DBTCONB.bit.DBT = DeadBandFine;
 }

}
```

Program Code #3

```
// Creating a Sine Modulated PWM Signal

// Include required header files
#include "DSP28_Device.h"
#include "volume.h"

int inp_buffer[BUFSIZE];       /* processing data buffers */
int i, j, value;

static void dataIO();   // dummy function to be used with ProbePoint

void main(void)
{
// Initialize System Control registers, PLL, WatchDog, Clocks to default state:
        InitSysCtrl();

// For this example, set HISPCLK to SYSCLKOUT/6 (25Mhz assuming 150Mhz
SYSCLKOUT)
  EALLOW;
  SysCtrlRegs.HISPCP.all = 0x3;    // HISPCLK = SYSCLKOUT/6
  EDIS;

        // Disable and clear all CPU interrupts:
        DINT;
        IER = 0x0000;
        IFR = 0x0000;

        // Initialize Pie Control Registers To Default State:
   InitPieCtrl();

        // Initialize the PIE Vector Table To a Known State:
        InitPieVectTable();
// Initialize the ADC
   InitAdc();   // DSP28_Adc.c

// Initialize EVB Timer3
        EvbRegs.T3PR = 0x5555;     // Timer3 period
        EvbRegs.T3CMPR = 0x3C00;         // Timer3 compare
        EvbRegs.T3CNT =  0x0000;         // Timer3 counter
        EvbRegs.T3CON.all  = 0x1042;       // Timer enable

// Enable compare
        EvbRegs.CMPR5 = 1250;

   // Compare action control.
   EvbRegs.ACTRB.all = 0x0666;
        EvbRegs.DBTCONB.bit.EDBT2 = 0;         // Disable deadband
   EvbRegs.COMCONB.all = 0xA600;
```

```
        i=0;

        dataIO();        // Read input data using a probe-point connected to a host file.

  while (1) // Infinite loop.
  {
  EALLOW; // Enable PWM pins.
  // Turn on/off the PWM signal through switch #2:
  GpioMuxRegs.GPBMUX.bit.PWM8_GPIOB1 = (GpioDataRegs.GPADAT.bit.GPIOA6+1);
  EDIS; // Disable PWM pins.

 value = (inp_buffer[i]+11000);
                j=0;
                while (j<200)
                {
                EvbRegs.CMPR4 = value;
                j++;
                }

      if (i<36)
      i++;
      else i=1;
  }

}

// ============ dataIO ============
//          read input signal
static void dataIO()
{
  /* do data I/O */
  return;
}
```

Program Code #4

```
// Program a

#include "volume.h"
#include "DSP28_Device.h"

int inp_buffer[BUFSIZE];       /* processing data buffers (in volume.h) */
int i, j, value, a, c, freq;
float b, mod;

interrupt void adc_isr(void);

Uint16 ConversionCount;
Uint16 Voltage1[10],Voltage2[10];

static void dataIO();   // dummy function to be used with ProbePoint

void main(void)
{
 // Initialize System Control registers, PLL, WatchDog, Clocks to default state:
        InitSysCtrl();

// For this example, set HISPCLK to SYSCLKOUT / 6 (25Mhz assuming 150Mhz
// SYSCLKOUT)
  EALLOW;
  SysCtrlRegs.HISPCP.all = 0x3;    // HISPCLK = SYSCLKOUT/6
  EDIS;

        // Disable and clear all CPU interrupts:
        DINT;
        IER = 0x0000;
        IFR = 0x0000;

        // Initialize Pie Control Registers To Default State:
        InitPieCtrl();

        // Initialize the PIE Vector Table To a Known State:
        InitPieVectTable();

   // Init the ADC
   InitAdc();   // DSP28_Adc.c

        EALLOW;    // This is needed to write to EALLOW protected registers
        PieVectTable.ADCINT = &adc_isr;
        EDIS;      // This is needed to disable write to EALLOW protected registers

   // Enable ADCINT in PIE
   PieCtrlRegs.PIEIER1.bit.INTx6 = 1;
```

```
    IER |= M_INT1;    // Enable Global INT1

    // Enable global Interrupts and higher priority real-time debug events:
        EINT;   // Enable Global interrupt INTM
        ERTM;           // Enable Global realtime interrupt DBGM

    ConversionCount = 0;

    // Configure ADC
 AdcRegs.ADCMAXCONV.all = 0x0001;       // Setup 2 conv's on SEQ1
 AdcRegs.ADCCHSELSEQ1.bit.CONV00 = 0x7; // Setup ADCINA7 as 1st SEQ1 conv.
 AdcRegs.ADCCHSELSEQ1.bit.CONV01 = 0x2; // Setup ADCINA2 as 2nd SEQ1 conv.
 AdcRegs.ADCTRL2.bit.EVA_SOC_SEQ1 = 1;  // Enable EVASOC to start SEQ1
 AdcRegs.ADCTRL2.bit.INT_ENA_SEQ1 = 1;  // Enable SEQ1 interrupt (every EOS)

 EvaRegs.T1CMPR = 0x0080;            // Setup T1 compare value
 EvaRegs.T1PR = 0xFFFF;              // Setup period register
 EvaRegs.GPTCONA.bit.T1TOADC = 1;       // Enable EVASOC in EVA

 EvaRegs.T1CON.all = 0x1042;          // Enable timer 1 compare (upcount mode)

                EvbRegs.T3CNT = 0x0000;      // Timer3 counter

        // Timer enable
        EvbRegs.T3CON.all = 0x1042;

    EvbRegs.ACTRB.all = 0x0666;
        EvbRegs.DBTCONB.all = 0x0000;

        EvbRegs.DBTCONB.bit.EDBT2=1; // Enable deadband
    EvbRegs.COMCONB.all = 0xA600;

     i=0;

// Enable PWM7 and PWM8:
EALLOW;
GpioMuxRegs.GPBMUX.bit.PWM8_GPIOB1 = 1;
GpioMuxRegs.GPBMUX.bit.PWM7_GPIOB0 = 1;
EDIS;

      dataIO(); // Read data from file

// set dead-time: 0.5 µs
EvbRegs.DBTCONB.bit.DBTPS = 1;
EvbRegs.DBTCONB.bit.DBT = 7;
```

```
while(1)
  {

// DIP#1=0, DIP#2=0
if ((GpioDataRegs.GPADAT.bit.GPIOA6==1)&&(GpioDataRegs.GPADAT.bit.GPIOA11== 1))
{
EvbRegs.T3PR = 10312; // set switching frequency
a=3728;
b=0.472;
}

// DIP#1=0, DIP#2=1
if ((GpioDataRegs.GPADAT.bit.GPIOA6==1)&&(GpioDataRegs.GPADAT.bit.GPIOA11==0))
{
EvbRegs.T3PR = 5156; // set switching frequency
a=1864;
b=0.236;
}

// DIP#1=1, DIP#2=0
if ((GpioDataRegs.GPADAT.bit.GPIOA6==0)&&(GpioDataRegs.GPADAT.bit.GPIOA11==1))
{
EvbRegs.T3PR = 2578; // set switching frequency
a=932;
b=0.118;
}

// DIP#1=1, DIP#2=1
if ((GpioDataRegs.GPADAT.bit.GPIOA6==0)&&(GpioDataRegs.GPADAT.bit.GPIOA11==0))
{
EvbRegs.T3PR = 1289; // set switching frequency
a=466;
b=0.059;
}

freq=(AdcRegs.ADCRESULT0>>13); // Truncate the least significant 13 bits
freq += 1; // sine frequency coefficient

mod=(AdcRegs.ADCRESULT1>>12); // Truncate the least significant 12 bits
mod = mod/15 + 0.0667; // modulation coefficient

value=(inp_buffer[i]+11000)*b*mod; // change sine values according to modulation and sine
                                   // frequency coefficients

            j=0;
            while (j<a)
            {
      EvbRegs.CMPR4 = value; // set duty cycle
      j++;
      }
```

```
      if (i<120)
      {
      i++;
     while (i%freq != 0)
     i++;
      }

      if ((i==120) || (i > 120))
      i=0;
    }
}

static void dataIO()
{
  /* do data I/O */
  return;
}

interrupt void  adc_isr(void)
{

  Voltage1[ConversionCount] = AdcRegs.ADCRESULT0;
  Voltage2[ConversionCount] = AdcRegs.ADCRESULT1;
  // If 10 conversions have been logged, start over
  if(ConversionCount == 9)
  {
    ConversionCount = 0;
  }
  else ConversionCount++;

  // Reinitialize for next ADC sequence
  AdcRegs.ADCTRL2.bit.RST_SEQ1 = 1;         // Reset SEQ1
  AdcRegs.ADCST.bit.INT_SEQ1_CLR = 1;                // Clear INT SEQ1 bit
  PieCtrlRegs.PIEACK.all = PIEACK_GROUP1;   // Acknowledge interrupt to PIE
  return;
}
```

Program Code #5

```
// Program b

#include "DSP28_Device.h"

// Prototype statements for functions found within this file.
interrupt void adc_isr(void);

// Global variables used in this example:

Uint16 ConversionCount;
Uint16 Voltage1[10], Voltage2[10];

float tastgrad;

void main(void)
{

// Step 1. Initialize System Control registers, PLL, WatchDog, Clocks to default state:
        InitSysCtrl();

// For this example, set HSPCLK to SYSCLKOUT / 1 and PLLCR register to 2
EALLOW;
  SysCtrlRegs.HISPCP.all = 1;
  SysCtrlRegs.PLLCR = 2;
EDIS;

        // Disable and clear all CPU interrupts:
        DINT;
        IER = 0x0000;
        IFR = 0x0000;

        // Initialize Pie Control Registers To Default State:
        InitPieCtrl();

        // Initialize the PIE Vector Table To a Known State:
        InitPieVectTable();

    // Init the ADC
    InitAdc();    // DSP28_Adc.c


        EALLOW;     // This is needed to write to EALLOW protected registers
        PieVectTable.ADCINT = &adc_isr;
        EDIS;       // This is needed to disable write to EALLOW protected registers

    // Enable ADCINT in PIE
    PieCtrlRegs.PIEIER1.bit.INTx6 = 1;
```

```
    IER |= M_INT1;    // Enable Global INT1

    // Enable global Interrupts and higher priority real-time debug events:
        EINT;  // Enable Global interrupt INTM
        ERTM;        // Enable Global realtime interrupt DBGM

    ConversionCount = 0;

    // Configure ADC
    AdcRegs.ADCMAXCONV.all = 0x0001;      // Setup 2 conv's on SEQ1
    AdcRegs.ADCCHSELSEQ1.bit.CONV00 = 0x7; // Setup ADCINA7 as 1st SEQ1 conv.
    AdcRegs.ADCCHSELSEQ1.bit.CONV01 = 0x2; // Setup ADCINA2 as 2nd SEQ1 conv.
    AdcRegs.ADCTRL2.bit.EVA_SOC_SEQ1 = 1; // Enable EVASOC to start SEQ1
    AdcRegs.ADCTRL2.bit.INT_ENA_SEQ1 = 1; // Enable SEQ1 interrupt (every EOS)

    EvaRegs.T1CMPR = 0x0080;          // Setup T1 compare value
    EvaRegs.T1PR = 0xFFFF;            // Setup period register
    EvaRegs.GPTCONA.bit.T1TOADC = 1;     // Enable EVASOC in EVA
    EvaRegs.T1CON.all = 0x1042;        // Enable timer 1 compare (upcount mode)

    // Initialize EVB Timer3
        EvbRegs.T3CMPR = 0x3C00;    // Timer3 compare
        EvbRegs.T3CNT = 0x0000;     // Timer3 counter
        // Timer enable
        EvbRegs.T3CON.all = 0x1042;

// Setup T3PWM and T4PWM
        // Drive T3/T4 PWM by compare logic
        EvbRegs.GPTCONB.bit.TCOMPOE = 1;
    // Polarity of GP Timer 3 Compare = Active low
        EvbRegs.GPTCONB.bit.T3PIN = 1;

    EvbRegs.ACTRB.all = 0x0666;
        EvbRegs.DBTCONB.all = 0x0000;
        EvbRegs.DBTCONB.bit.EDBT2=1; // Enable deadband
    EvbRegs.COMCONB.all = 0xA600;

// Enable PWM9 and PWM10
EALLOW;
GpioMuxRegs.GPBMUX.bit.PWM10_GPIOB3 = 1;
GpioMuxRegs.GPBMUX.bit.PWM9_GPIOB2 = 1;
EDIS;

    while (1)
    {
// DIP#1=0, DIP#2=0: dead-time = 0.5 µs
if ((GpioDataRegs.GPADAT.bit.GPIOA6==1)&&(GpioDataRegs.GPADAT.bit.GPIOA11==1))
{
EvbRegs.DBTCONB.bit.DBTPS = 1;
EvbRegs.DBTCONB.bit.DBT = 4;
}
```

```
// DIP#1=0, DIP#2=1: dead-time = 1 µs
if ((GpioDataRegs.GPADAT.bit.GPIOA6==1)&&(GpioDataRegs.GPADAT.bit.GPIOA11==0))
{
EvbRegs.DBTCONB.bit.DBTPS = 2;
EvbRegs.DBTCONB.bit.DBT = 4;
}

// DIP#1=1, DIP#2=0: dead-time = 2 µs
if ((GpioDataRegs.GPADAT.bit.GPIOA6==0)&&(GpioDataRegs.GPADAT.bit.GPIOA11==1))
{
EvbRegs.DBTCONB.bit.DBTPS = 3;
EvbRegs.DBTCONB.bit.DBT = 4;
}

// DIP#1=1, DIP#2=1: dead-time = 3 µs
if ((GpioDataRegs.GPADAT.bit.GPIOA6==0)&&(GpioDataRegs.GPADAT.bit.GPIOA11==0))
{
EvbRegs.DBTCONB.bit.DBTPS = 3;
EvbRegs.DBTCONB.bit.DBT = 6;
}

   EvbRegs.T3PR=((AdcRegs.ADCRESULT0>>12) + 1)*125;
   tastgrad=(AdcRegs.ADCRESULT1>>11)*3.4 + 0.7;
   EvbRegs.CMPR5 = EvbRegs.T3PR*(tastgrad/100);

   }

}

interrupt void adc_isr(void)
{

  Voltage1[ConversionCount] = AdcRegs.ADCRESULT0;
  Voltage2[ConversionCount] = AdcRegs.ADCRESULT1;
  // If 10 conversions have been logged, start over
  if (ConversionCount == 9)
  {
    ConversionCount = 0;
  }
  else ConversionCount++;

  // Reinitialize for next ADC sequence
  AdcRegs.ADCTRL2.bit.RST_SEQ1 = 1;         // Reset SEQ1
  AdcRegs.ADCST.bit.INT_SEQ1_CLR = 1;       // Clear INT SEQ1 bit
   PieCtrlRegs.PIEACK.all = PIEACK_GROUP1;  // Acknowledge interrupt to PIE
   return;
}
```

Program Code #6

```
// Program c

#include "DSP28_Device.h"
#include "math.h"

// Prototype statements for functions found within this file:
interrupt void adc_isr(void);

// Global variables used in this example:

Uint16 ConversionCount;
Uint16 Voltage1[50];
Uint16 Voltage2[50];

float a, b, c, d, p, average, result;
float value, frequency;
const float pi=3.14;
int ok, z, freq_int;
int freq, freq2;

main()
{
InitSysCtrl();

EALLOW;
SysCtrlRegs.HISPCP.all = 0x1;       // HSPCLK = SYSCLKOUT/6
SysCtrlRegs.PLLCR = 2;
EDIS;

        DINT;
        IER = 0x0000;
        IFR = 0x0000;

        InitPieCtrl();
        InitPieVectTable();

   InitAdc();

        EALLOW;
        PieVectTable.ADCINT = &adc_isr;
        EDIS;

   PieCtrlRegs.PIEIER1.bit.INTx6 = 1;
   IER |= M_INT1;

EINT;
ERTM;
```

```
   ConversionCount = 0;

   // Configure ADC
   AdcRegs.ADCMAXCONV.all = 0x0001;      // Setup 2 conv's on SEQ1
   AdcRegs.ADCCHSELSEQ1.bit.CONV00 = 0x0; // Setup ADCINA0 as 1st SEQ1 conv.
   AdcRegs.ADCCHSELSEQ1.bit.CONV01 = 0x7; // Setup ADCINA7 as 2nd SEQ1 conv.
   AdcRegs.ADCTRL2.bit.EVA_SOC_SEQ1 = 1;  // Enable EVASOC to start SEQ1
   AdcRegs.ADCTRL2.bit.INT_ENA_SEQ1 = 1;  // Enable SEQ1 interrupt (every EOS)

   // Configure EVA
   EvaRegs.T1PR = 75;            // Setup period register
   EvaRegs.T1CMPR = 35;          // Setup T1 compare value
   EvaRegs.GPTCONA.bit.T1TOADC = 1;      // Enable EVASOC in EVA

// Timer compare enable
        EvaRegs.T1CON.all = 0x1042;

        EvbRegs.T3PR = 750;
        EvbRegs.T3CON.all = EvaRegs.T1CON.all = 0x1042;

   EvbRegs.ACTRB.all = 0x0666;
   EvbRegs.COMCONB.all = 0xA600;

EvbRegs.CMPR5 = 375;

ok=0;
p = 0;
value = 0;
frequency = 0;
average=0;
result=1;
freq=freq2=750;

EvbRegs.DBTCONB.bit.EDBT2=1; // Enable Deadband

// Set Dead-Time: 0.5µs
EvbRegs.DBTCONB.bit.DBTPS = 1;
EvbRegs.DBTCONB.bit.DBT = 4;

EALLOW;
GpioMuxRegs.GPBMUX.bit.PWM10_GPIOB3 = 1;
GpioMuxRegs.GPBMUX.bit.PWM9_GPIOB2 = 1;
GpioMuxRegs.GPAMUX.bit.CAP2Q2_GPIOA9=0;
GpioMuxRegs.GPAMUX.bit.T1PWM_GPIOA6=0;
EDIS;

EvbRegs.T3CNT = 0;
EvaRegs.T1CNT = 1;

       while (1)
   {
```

```
// If DIP#1=0;
if ((GpioDataRegs.GPADAT.bit.GPIOA6==1)&&(GpioDataRegs.GPADAT.bit.GPIOA11==1))
{
// Set frequency according to the analog input (potentiometer)
freq=590 + (AdcRegs.ADCRESULT1>>12)*27;
EvbRegs.T3PR=freq;
EvbRegs.CMPR5 = freq/2;
}

// If DIP#1=1;
if ((GpioDataRegs.GPADAT.bit.GPIOA6==0)&&(GpioDataRegs.GPADAT.bit.GPIOA11==1))
{
// Set frequency according to the calculated values
if ((frequency != 0) && (frequency > 5000) && (frequency < 30000))
{
freq_int=frequency;
freq2 = (750/((freq_int>>10)+1))*20;
freq2 = freq2*(1.09) - 56;
}
EvbRegs.T3PR = freq2;
EvbRegs.CMPR5 = EvbRegs.T3PR/2;
}

if (result < 1)
        {
        value = acos(result);
        frequency = (value*200000)/(2*pi);
        }
}

}

interrupt void adc_isr(void)
{

if (ok == 0)
  {
  Voltage1[ConversionCount] = AdcRegs.ADCRESULT0;

  // If 50 conversions have been logged, start over
  if (ConversionCount > 48)
  {
  ConversionCount = 0;
  ok = 1;
  }
  else ConversionCount++;
  }

while (ok==1)
{
```

```
if (z < 47)
{
a = Voltage1[z];
b = Voltage1[z+1];
c = Voltage1[z+2];
d = Voltage1[z+3];
}

if ((a > 38000) && (b > 38000) && (c > 38000) && (d > 38000))
{
z++;

a= ((3.5/26000)*a - 5.1);
b= ((3.5/26000)*b - 5.1);
c= ((3.5/26000)*c - 5.1);
d= ((3.5/26000)*d - 5.1);
p = (p + ((a + c)*b + c*(b + d))/(2*(b*b + c*c)));
average++;

if (average == 50)
   {
   result = p/average;
   p=0;
   average=0;
   }
}

else
z++;

if (z>46)
{
z=0;
ok = 0;
}

}

  // Reinitialize for next ADC sequence
  AdcRegs.ADCTRL2.bit.RST_SEQ1 = 1;         // Reset SEQ1
  AdcRegs.ADCST.bit.INT_SEQ1_CLR = 1;                    // Clear INT SEQ1 bit
  PieCtrlRegs.PIEACK.all = PIEACK_GROUP1;   // Acknowledge interrupt to PIE

  return;
}
```

B. Circuitry and Wiring Diagram of the Experimental Set-Up

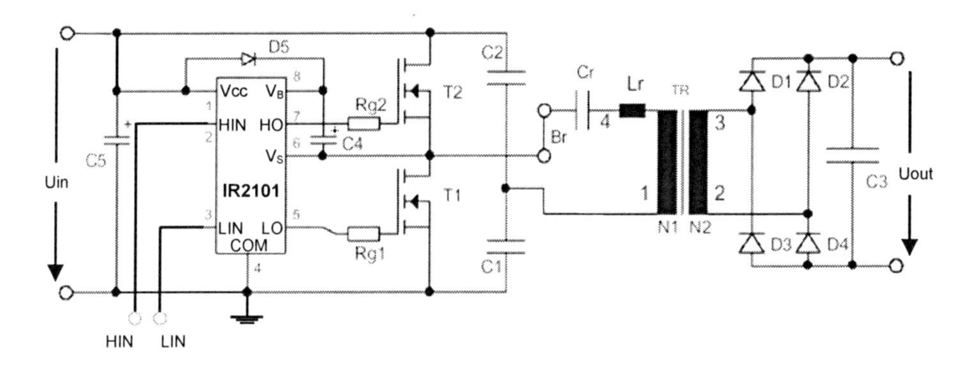

Figure B.1 Circuit Used in Control of SRC

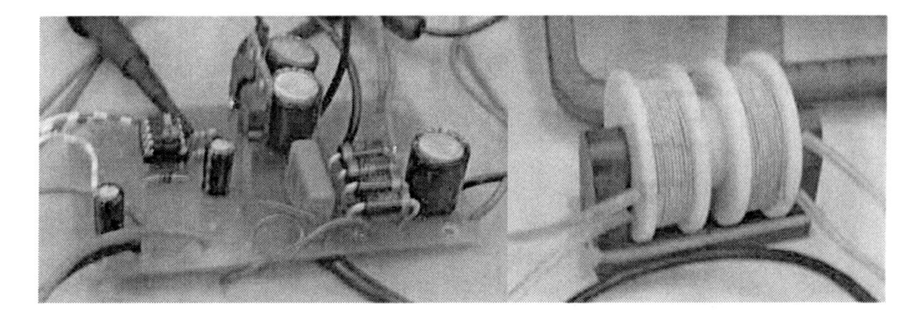

Figure B.2 Photograph of the SRC Set-Up

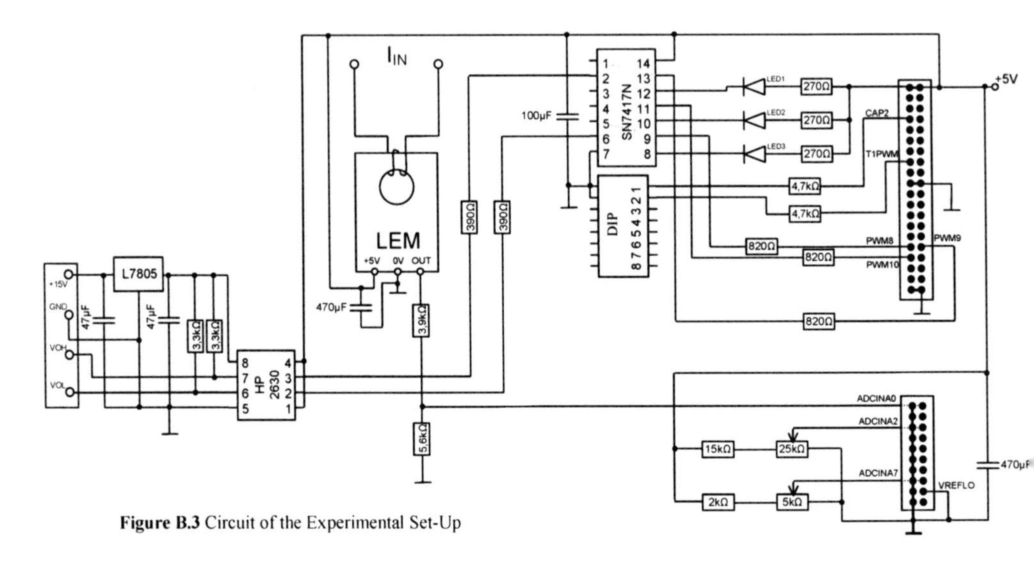

Figure B.3 Circuit of the Experimental Set-Up

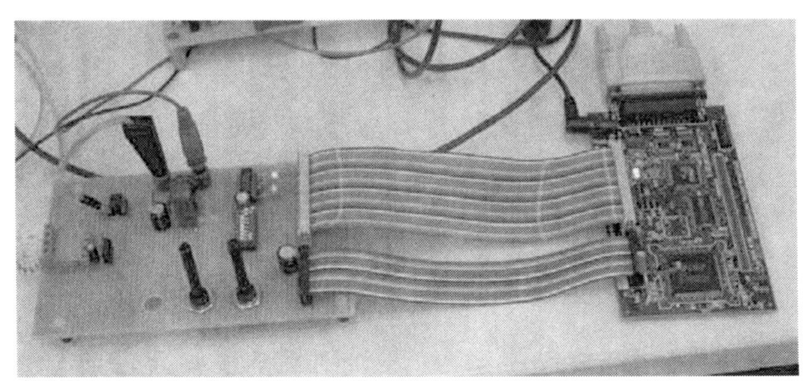

Figure B.4 Photograph of the Experimental Set-Up

C. Acronyms and Abbreviations

AC	Alternating Current
ADC	Analog-to-Digital Converter
ALU	Arithmetic Logic Unit
ASIC	Application Specific Integrated Circuits
CCS	Code Composer Studio
COFF	Common Object File Format
CSI	Current Source Inverter
DAC	Digital-to-Analog Converter
DC	Direct Current
DCS	Digital Control System
DIP	Dual In Line Package
DMA	Direct Memory Access
DSP	Digital Signal Processor
eCAN	Enhanced Controller Area Network
ED	Electrical Drive
EMI	Electromagnetic Interference
EOS	End-of-Sequence
EPROM	Erasable Programmable Read Only Memory
EV	Event Manager
EVM	Event Manager Module
FFT	Fast Fourier Transform
FIFO	First-In First-Out buffer
FIR	Finite Impulse Response
GP	General Purpose
GPIO	General Purpose Input Output
GPP	General Purpose Processor
HF	High Frequency
I/O	Input/Output
JTAG	Joint Test Action Group
LED	Light Emitting Diode
MAC	Multiply and Accumulate Operation
McBSP	Multichannel Buffered Serial Port

MIPS	Millions of Instructions Per Second
OI	Output Interface
PE	Power Electronic
PID	Proportional-Integral-Derivative
PIE	Peripheral Interrupt Expansion
PWM	Pulse Width Modulation
QEP	Quadrature-Encoder Pulse
RAM	Random Access Memory
ROM	Read Only Memory
SARAM	Single Access RAM
SCI	Serial Communications Interface
SPI	Serial Peripheral Interface
SRC	Series Resonant Converter
TI	Texas Instruments
UPS	Uninterruptible Power Supply
VRM	Voltage Regulator Module
VSI	Voltage Source Inverter
XINTF	External Interface
ZCS	Zero Current Switching
ZVS	Zero Voltage Switching

D. Schematics of the eZdsp F2812 Board

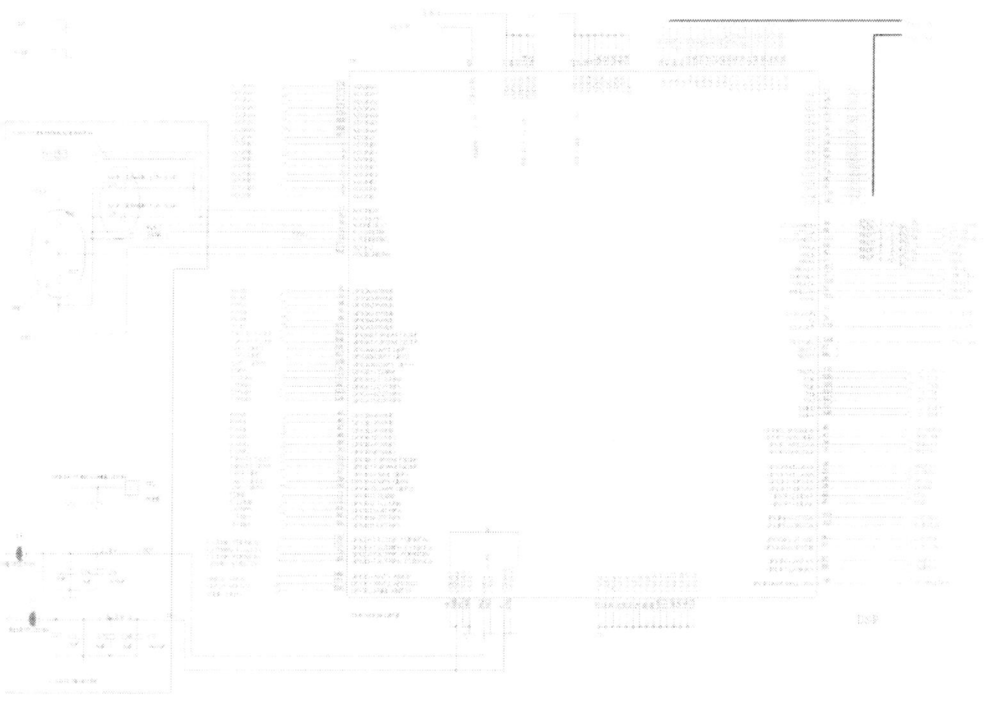

Figure D.1 Schematic of the DSP [55]

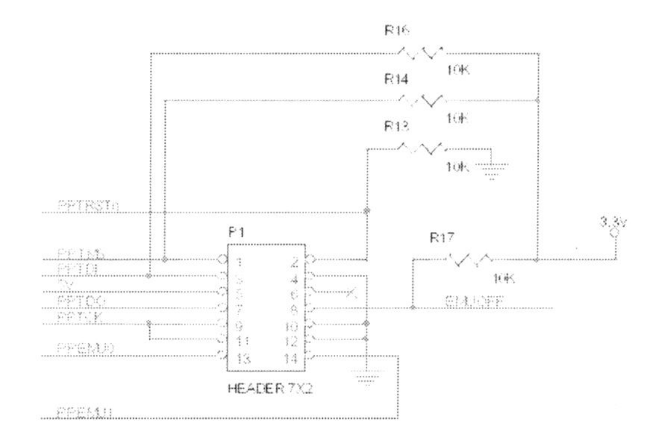

Figure D.2 Schematic of the P1 Header [55]

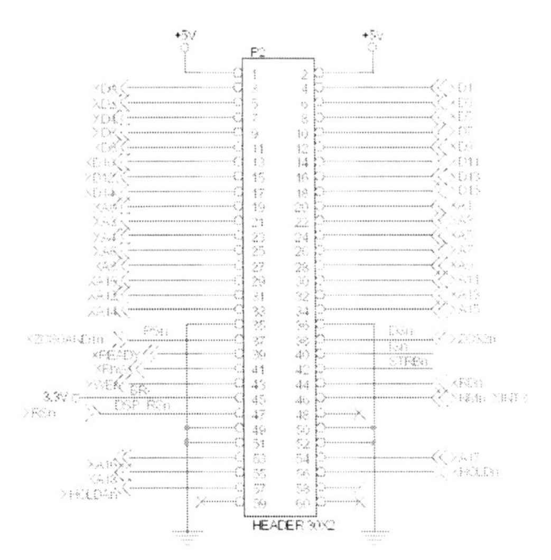

Figure D.3 Schematic of the P2 Header [55]

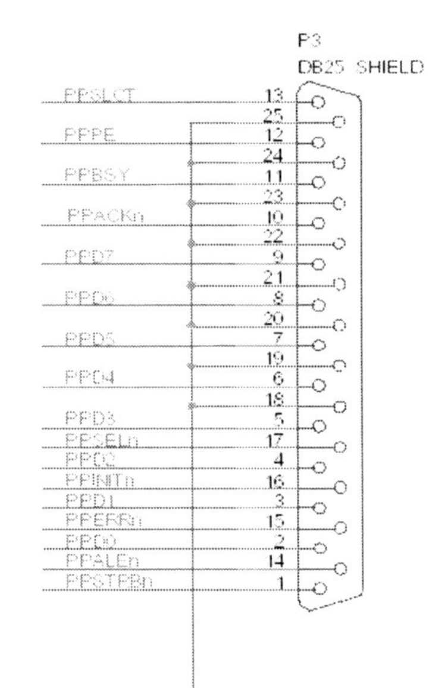

Figure D.4 Schematic of the P3 Parallel Interface [55]

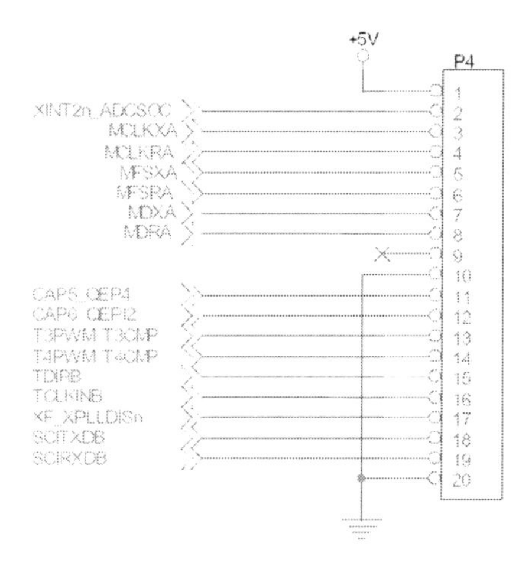

Figure D.5 Schematic of the P4 Header [55]

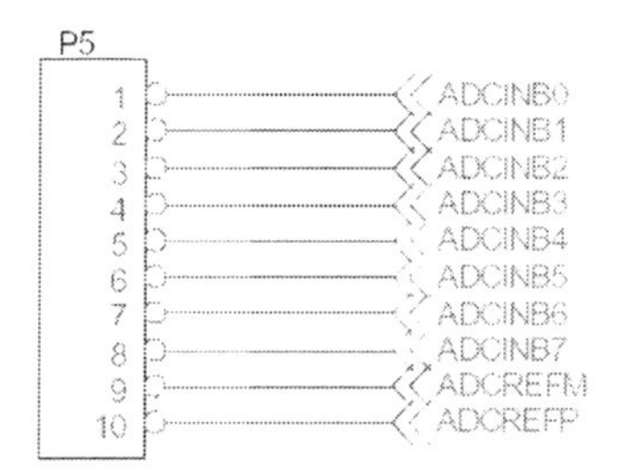

Figure D.6 Schematic of the P5 Header [55]

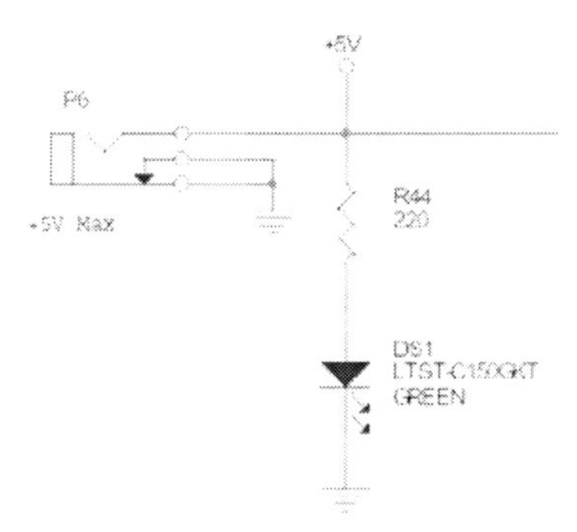

Figure D.7 Schematic of the P6 Power Connector [55]

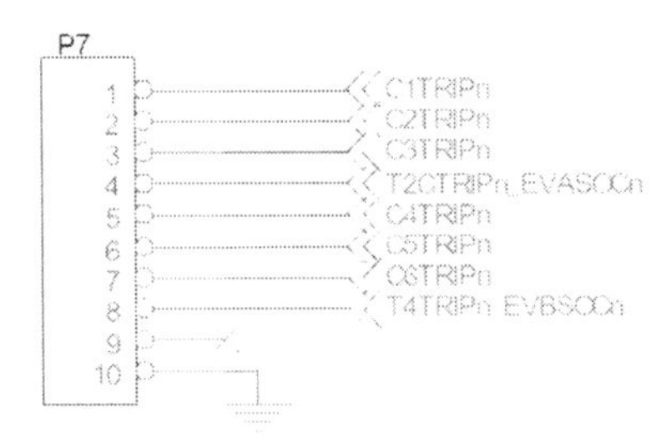

Figure D.8 Schematic of the P7 Header [55]

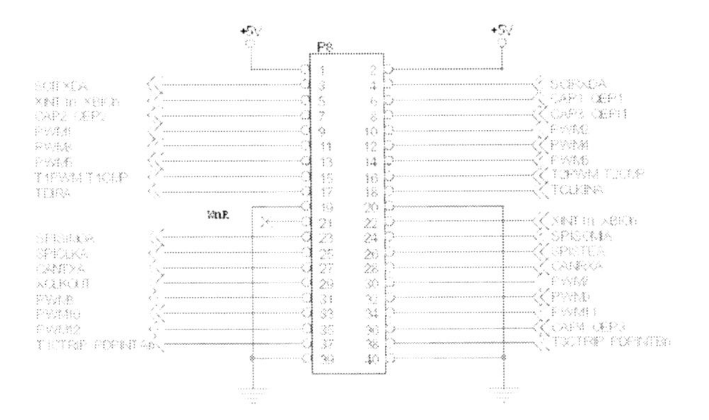

Figure D.9 Schematic of the P8 Header [55]

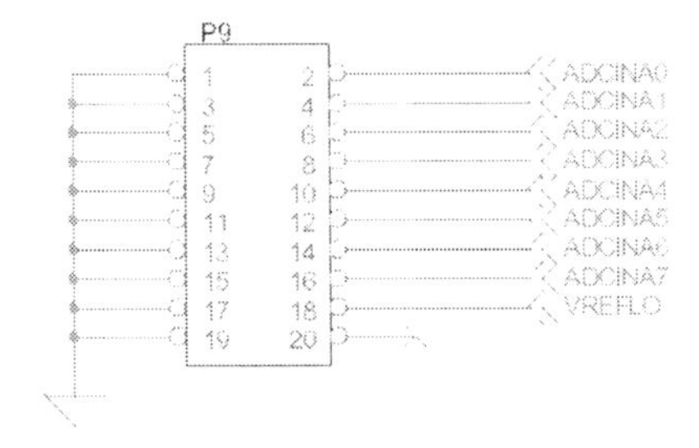

Figure D.10 Schematic of the P9 Header [55]

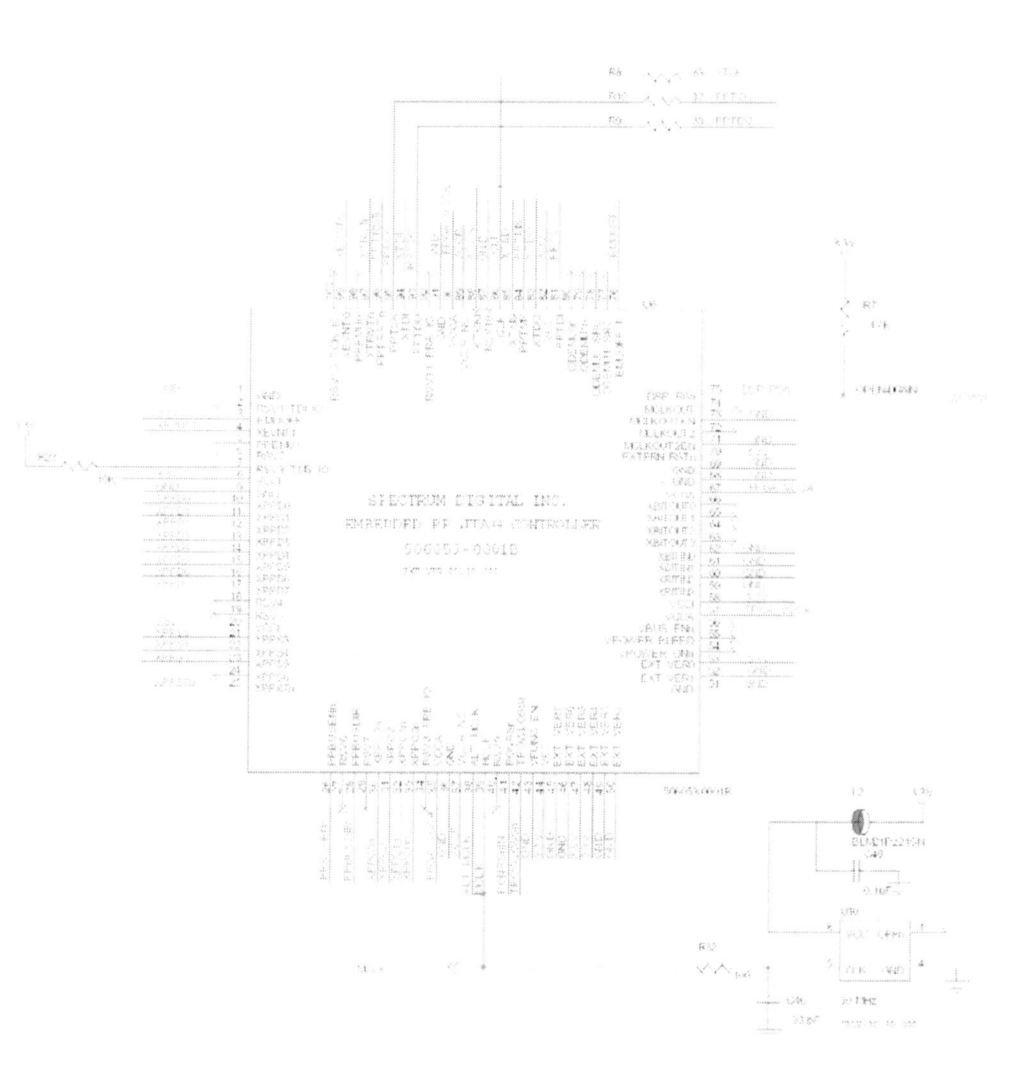

Figure D.11 Schematic of the JTAG Controller [55]

E. Sine Values Contained in sinus.dat

1651 1 0 1 0 (This must be the first line in order to make the file readable from CCS)

1	0	31	10000	61	0	91	-10000
2	523	32	9986	62	-523	92	-9986
3	1045	33	9945	63	-1045	93	-9945
4	1564	34	9877	64	-1564	94	-9877
5	2079	35	9781	65	-2079	95	-9781
6	2588	36	9659	66	-2588	96	-9659
7	3090	37	9511	67	-3090	97	-9511
8	3584	38	9336	68	-3584	98	-9336
9	4067	39	9135	69	-4067	99	-9135
10	4540	40	8910	70	-4540	100	-8910
11	5000	41	8660	71	-5000	101	-8660
12	5446	42	8387	72	-5446	102	-8387
13	5878	43	8090	73	-5878	103	-8090
14	6293	44	7771	74	-6293	104	-7771
15	6691	45	7431	75	-6691	105	-7431
16	7071	46	7071	76	-7071	106	-7071
17	7431	47	6691	77	-7431	107	-6691
18	7771	48	6293	78	-7771	108	-6293
19	8090	49	5878	79	-8090	109	-5878
20	8387	50	5446	80	-8387	110	-5446
21	8660	51	5000	81	-8660	111	-5000
22	8910	52	4540	82	-8910	112	-4540
23	9135	53	4067	83	-9135	113	-4067
24	9336	54	3584	84	-9336	114	-3584
25	9511	55	3090	85	-9511	115	-3090
26	9659	56	2588	86	-9659	116	-2588
27	9781	57	2079	87	-9781	117	-2079
28	9877	58	1564	88	-9877	118	-1564
29	9945	59	1045	89	-9945	119	-1045
30	9986	60	523	90	-9986	120	-523

Table E.1 The Sine Values Stored in File

CPSIA information can be obtained at www.ICGtesting.com
Printed in the USA
BVOW03s2013080814

362208BV00001B/7/P

9 783638 701846